Nonlinear Equations and Operator Algebras

Mathematics and Its Applications (*Soviet Series*)

Vladimir A. Marchenko

Low Temperature Institute, Academy of Sciences,
Kharkov, Ukrainian SSR

Nonlinear Equations and Operator Algebras

Translated by V. I. Rublinetsky

D. Reidel Publishing Company

A MEMBER OF THE KLUWER ACADEMIC PUBLISHERS GROUP

Dordrecht / Boston / Lancaster / Tokyo

Library of Congress Cataloging in Publication Data

Marchenko, V. A. (Vladimir Aleksandrovich), 1922–
[Nelineĭnye uravneniia i operatornye algebry. English]
Nonlinear equations and operator algebras / by Vladimir A. Marchenko; translated by V. I. Rublinetsky.
p. cm. — (Mathematics and its applications (Soviet series))
Translation of: Nelineĭnye uravneniia i operatornye algebry.
Includes bibliographies and index.
ISBN 90-277-2654-X
1. Differential equations, Nonlinear. 2. Operator algebras. I. Title.
II. Series: Mathematics and its applications (D. Reidel Publishing Company. Soviet series)
QA372.M36513 1987
515.3′55—dc 19 87-28906
CIP

Published by D. Reidel Publishing Company,
P.O. Box 17, 3300 AA Dordrecht, Holland.

Sold and distributed in the U.S.A. and Canada
by Kluwer Academic Publishers,
101 Philip Drive, Norwell, MA 02061, U.S.A.

In all other countries, sold and distributed
by Kluwer Academic Publishers Group,
P.O. Box 322, 3300 AH Dordrecht, Holland.

Original Russian edition published by Naukova Dumka

Printed in The Netherlands

SERIES EDITOR'S PREFACE

Approach your problems from the right end and begin with the answers. Then one day, perhaps you will find the final question.

'The Hermit Clad in Crane Feathers' in R. van Gulik's *The Chinese Maze Murders*.

It isn't that they can't see the solution. It is that they can't see the problem.

G.K. Chesterton. *The Scandal of Father Brown* 'The point of a Pin'.

Growing specialization and diversification have brought a host of monographs and textbooks on increasingly specialized topics. However, the "tree" of knowledge of mathematics and related fields does not grow only by putting forth new branches. It also happens, quite often in fact, that branches which were thought to be completely disparate are suddenly seen to be related.

Further, the kind and level of sophistication of mathematics applied in various sciences has changed drastically in recent years: measure theory is used (non-trivially) in regional and theoretical economics; algebraic geometry interacts with physics; the Minkowsky lemma, coding theory and the structure of water meet one another in packing and covering theory; quantum fields, crystal defects and mathematical programming profit from homotopy theory; Lie algebras are relevant to filtering; and prediction and electrical engineering can use Stein spaces. And in addition to this there are such new emerging subdisciplines as "experimental mathematics", "CFD", "completely integrable systems", "chaos, synergetics and large-scale order", which are almost impossible to fit into the existing classification schemes. They draw upon widely different sections of mathematics. This programme, Mathematics and Its Applications, is devoted to new emerging (sub)disciplines and to such (new) interrelations as exempla gratia:

- a central concept which plays an important role in several different mathematical and/or scientific specialized areas;
- new applications of the results and ideas from one area of scientific endeavour into another;
- influences which the results, problems and concepts of one field of enquiry have and have had on the development of another.

The Mathematics and Its Applications programme tries to make available a careful selection of books which fit the philosophy outlined above. With such books, which are stimulating rather than definitive, intriguing rather than encyclopaedic, we hope to contribute something towards better communication among the practitioners in diversified fields.

Because of the wealth of scholarly research being undertaken in the Soviet Union, Eastern Europe, and Japan, it was decided to devote special attention to the work emanating from these particular regions. Thus it was decided to start three regional series under the umbrella of the main MIA programme.

It is probably by now largely superfluous to write about the importance of the so-called soliton equations such as the Korteweg-de Vries equation, the cubic nonlinear Schrödinger equation, and so many others. They are important in terms of direct applications, in terms of model equations for physical theories (such as quantum field theory) and in terms of the interrelations with various parts of mathematics such as representation theory and algebraic geometry.

One of the astonishing, and so far not understood, aspects of these equations is the variety of different ways which exist to obtain their solutions, and the relations between these various approaches.

This book presents one more way of finding solutions of clearly very considerable power and about which it is still largely an open question just how powerful it will turn out to be. It starts with the not unknown observation that the fractional linear transformation $\Gamma \mapsto \Gamma^{-1}\Gamma_x$ takes the solutions of a linear set of equations to solutions of the nonlinear KdV equation. The next step consists of the couple of beautiful and far reaching ideas that first the same thing can be done for suitable (non commutative) operator algebras to give solutions of an operator KdV equation and that these solutions can then be projected to yield solutions of the original KdV, including solutions not accessible by the inverse scattering transform method and the algebraic-geometric Jacobian variety method.

Thus this book adds one more fascinating and promising chunk of theory and technique to the fascinating literature on integrable systems, and on the (much related) Riemann problem.

The unreasonable effectiveness of mathematics in science ...

Eugene Wigner

Well, if you know of a better 'ole, go to it.

Bruce Bairnsfather

What is now proved was once only imagined.

William Blake

As long as algebra and geometry proceeded along separate paths, their advance was slow and their applications limited.

But when these sciences joined company they drew from each other fresh vitality and thenceforward marched on at a rapid pace towards perfection.

Joseph Louis Lagrange.

Bussum, November 1987

Michiel Hazewinkel

CONTENTS

PREFACE

In 1967, Gardner, Green, Kruskal, and Miura [30] discovered a method for integrating the Korteweg–de Vries equation with the aid of the inverse problem of scattering theory. Further development of this method by Lax [15], Zakharov, and Shabat [7], [8], and other mathematicians and physicists made it possible to find solutions of a number of nonlinear equations important in physics, and to understand the connection of this method with the theory of Hamiltonian systems with an infinite number of degrees of freedom (Gardner [29], Zakharov, and Faddeev [6]). In addition, various modifications of the method were constructed (Zakharov and Shabat [8], Zakharov and Mikhailov [5]). In 1974–75, a method of finding periodic and almost periodic solutions of these equations was suggested (Novikov [21], Lax [31], Marchenko [16]).

Already in 1961, Akhiezer [1] discovered the connection between inverse problems for some Sturm–Liouville operators with a finite gap spectrum and the problem of inverting Jacobi–Abel integrals. Developing Akhiezer's idea, Its and Matveew [12], Dubrovin and Novikov [2], McKean and Van Moerbeke [33] found explicit formulas for finite zone solutions of the Korteweg–de Vries equation.

Soon similar results were obtained for a number of other nonlinear equations [10], [11], [13]. The authors substantially used algebraic geometry methods which were later developed by Krichever [14].

The inverse problem method and the methods connected with algebraic geometry were substantially presented in the monograph [4] by Zakharov, Manakov, Novikov, and Pitayevsky (the book contains a comprehensive bibliography, including important surveys and collections of articles).

Intensive consideration was also given to other approaches that enable finding integrable equations and their solutions (the Bäcklund transformation, the Hirota method [26], the Adler scheme [28], the Matveew method based on the Darboux–Krum transformation [32], the Fokas–Ablowitz method [27], and others).

The present book describes the method of integrating nonlinear equations suggested by the author in 1979. Some applications of this method were studied in articles [17], [22]–[24], and in the thesis by Tarapova [25]. Chapters 1 and 2 are partially devoted to these results.

While writing the book, I was much aided by I. V. Ostrovskiǐ, especially in questions concerning the Carleson and Muckenhoupt theorems. I wish to thank him heartily for his friendly help.

Kharkov, 1986

INTRODUCTION

This book describes a method for integrating nonlinear differential equations. The method is based upon the following observations on one-soliton solutions of the equation

$$u_t + 6u_x^2 + u_{xxx} = 0 \tag{1}$$

which is equivalent, as can be easily seen, to the Korteweg–de Vries (KdV) equation $v_t - 6vv_x + v_{xxx} = 0$ $(v = -2u_x)$.

1. Propagating waves $u = u(x - a^2 t)$ are the simplest solutions of Eq. (1). Such solutions are called *one-soliton* and can be easily found, since this case is reduced to solving an ordinary differential equation integrable in quadratures. Once the appropriate values for integration constants are chosen, the solutions obtained acquire the form

$$u = \Gamma^{-1}\Gamma_x, \qquad \Gamma = e^{ax-4a^3t} + e^{(-ax-4a^3t)}m, \tag{2}$$

where a and m are arbitrary numerical parameters.

Note that the function $\Gamma = \Gamma(x,t)$ satisfies the following linear differential equations with constant coefficients,

$$\Gamma_t + 4\Gamma_{xxx} = 0, \qquad \Gamma_{xx} = a^2\Gamma, \tag{3}$$

that determine Γ uniquely to within a numerical factor. The equation $\Gamma_t + 4\Gamma_{xxx} = 0$ is the linear part of Eq. (1), while its general solution can be represented as a superposition of waves propagating with different velocities. The second equation, $\Gamma_{xx} = a^2\Gamma$, separates among the waves the one which propagates with the fixed velocity a^2, i.e., it selects one-soliton solutions from the set of solutions of the first equation. Thus, we have arrived at the following statement:

One-soliton solutions of the nonlinear Eq. (1) are logarithmic derivatives $(u = \Gamma^{-1}\Gamma_x)$ of one-soliton solutions (Γ) of its linear part (the irrelevant numerical coefficient equal to 4 can be removed by rescaling).

2. Surely, one can give the direct proof of the above statement, checking immediately that the logarithmic derivative $u = \Gamma^{-1}\Gamma_x$ of any function satisfying Eq. (3) is a solution of Eq. (1). To do this, one has to carry out easy calculations that are, after all, nothing but standard arithmetic operations and differentiations. In differentiating, one uses only linearity and commutativity, as

well as the Leibniz formula for the product. Commutativity of multiplication is not used in these calculations, so the above statement is valid also for operator-valued solutions $\hat{u}(x,t)$ of Eq. (1) whose values belong to an arbitrary operator algebra, provided that commutative differential operators are defined correctly, so that these operators ∂_t, ∂_x possess the standard algebraic properties

$$\partial(\hat{u}+\hat{v}) = \partial(\hat{u}) + \partial(\hat{v}), \qquad \partial(\hat{u}\hat{v}) = \partial(\hat{u})\hat{v} + \hat{u}\partial(\hat{v}).$$

By analogy with the scalar case, operator functions $\hat{\gamma} = \hat{\Gamma}^{-1}(\partial\hat{\Gamma})$ will be called logarithmic derivatives of operator functions $\hat{\Gamma}$ with respect to differentiations ∂.

The above considerations show that in passing from Eq. (1) to the equation of similar form,

$$\hat{u}_t + 6\hat{u}_x^2 + \hat{u}_{xxx} = 0, \tag{4}$$

in operator-valued functions $\hat{u} = \hat{u}(x,t)$, one can confirm that the logarithmic derivatives $\hat{\gamma} = \widehat{\Gamma}^{-1}\widehat{\Gamma}_x$ of operator functions $\widehat{\Gamma} = \widehat{\Gamma}(x,t)$ that satisfy the equations

$$\widehat{\Gamma}_t + 4\widehat{\Gamma}_{xxx} = 0, \qquad \widehat{\Gamma}_{xx} = \hat{a}^2\widehat{\Gamma}, \tag{5}$$

where $\hat{a}$ stand for arbitrary constant operators, solve Eq. (4).

Eq. (2) is also valid in the operator case if constant numbers a and m are substituted by constant operators $\hat{a}$ and $\hat{m}$:

$$\widehat{\Gamma} = e^{\hat{a}x - 4\hat{a}^3 t} + e^{-(\hat{a}x - 4\hat{a}^3 t)}\hat{m}. \tag{6}$$

Operator functions $\widehat{\Gamma}$ then can naturally be called operator solitons of the equation $\widehat{\Gamma}_t + 4\widehat{\Gamma}_{xxx} = 0$, while their logarithmic derivatives $\hat{\gamma} = \widehat{\Gamma}^{-1}\widehat{\Gamma}_x$ are called operator solitons of Eq. (4).

3. Note now that if an operator function $\hat{u} = \hat{u}(x,t)$ satisfies Eq. (4), then functions $\hat{u}(x,t) + \hat{c}$, where $\hat{c}$ are arbitrary constant operators, also satisfy these equations. In other words, Eq. (4) admits an additive transformation group $\hat{u} \to \hat{u} + \hat{c}$ consisting of all constant operators $\hat{c}$.

Every solution $\hat{u}(x,t)$ of Eq. (4) can be decomposed into a sum of two terms

$$\hat{u}(x,t) = \hat{u}(x,t)\widehat{P} + \hat{u}(x,t)(I - \widehat{P}),$$

where $\widehat{P} = \widehat{P}^2$ is an arbitrarily chosen constant projection operator onto a one-dimensional space.

If in the above decomposition the second term is a constant operator $\widehat{N}$,

$$\hat{u}(x,t)(I - \widehat{P}) = \widehat{N}, \tag{7}$$

then the first term $\hat{u}(x,t)\widehat{P}$ will also satisfy Eq. (4),

$$\hat{u}_t\widehat{P} + 6\hat{u}_x\widehat{P}\hat{u}_x\widehat{P} + \hat{u}_{xxx}\widehat{P} = 0.$$

Multiplying both sides of this equation by $\hat{P}$ from the left, we obtain

$$(8) \qquad (\hat{P}\hat{u}\hat{P})_t + 6(\hat{P}\hat{u}\hat{P})_x(\hat{P}\hat{u}\hat{P})_x + (\hat{P}\hat{u}\hat{P})_{xxx} = 0,$$

since

$$\hat{P}\hat{u}_x\hat{P} = \hat{P}\hat{u}_x\hat{P}^2 = (\hat{P}\hat{u}\hat{P})_x\hat{P}.$$

Since every linear operator in a one-dimensional space is an operator of multiplication by a number, then $\hat{P}\hat{u}(x,t)\hat{P} = u(x,t)\hat{P}$, and Eq. (8) implies that the scalar function $u(x,t)$ satisfies the initial equation (1). The mapping $\hat{u}(x,t) \to u(x,t)$ defined by the equality $\hat{P}\hat{u}(x,t)\hat{P} = u(x,t)\hat{P}$ is called a *projection operation*, and the scalar function $u(x,t)$ is a one-dimensional projection of the operator function $\hat{u}(x,t)$.

Hence, one-dimensional projections of solutions of the operator Eq. (4) that satisfy condition (7) solve the initial Eq. (1).

4. It is clear that one-soliton solutions $\hat{\gamma} = \hat{\Gamma}^{-1}\hat{\Gamma}_x$ of the operator Eq. (4) satisfy condition (7) if and only if the operator functions $\hat{\Gamma} = \hat{\Gamma}(x,t)$ satisfy the equation

$$\hat{\Gamma}_x(I - \hat{P}) = \hat{\Gamma}\hat{N}(I - \hat{P}).$$

Associating the facts given in the above subsections, we arrive at the basic idea of the suggested method for integration of nonlinear equations.

Let an operator function $\hat{\Gamma} = \hat{\Gamma}(x,t)$ defined in a certain domain of the x,t-plane satisfy the equation

$$(9) \qquad \hat{\Gamma}_t + 4\hat{\Gamma}_{xxx} = 0, \qquad \hat{\Gamma}_{xx} = \hat{a}^2\hat{\Gamma}, \qquad \hat{\Gamma}_x(I - \hat{P}) = \hat{\Gamma}\hat{N}(I - \hat{P})$$

where $\hat{a}, \hat{N}$ are arbitrary constant operators, and $\hat{P} = \hat{P}^2$ is an arbitrary one-dimensional constant projector.

If all points of the above domain of the operators $\hat{\Gamma}(x,t)$ are invertible, then the logarithmic derivative $\hat{\gamma} = \hat{\gamma}(x,t) = \hat{\Gamma}^{-1}\hat{\Gamma}_x$ exists and satisfies the operator Eq. (4), while its one-dimensional projection $u(x,t)$ defined by the equality $\hat{P}\hat{u}(x,t)\hat{P} = u(x,t)\hat{P}$ satisfies the scalar Eq. (1).

Remark. In accordance with Eq. (7), one can find the operator functions $\hat{\Gamma}(x,t)$ in the form

$$\hat{\Gamma} = e^{\hat{a}x - 4\hat{a}^3 t}(I + \hat{T}(x,t)).$$

It can easily be seen that Eqs. (9) are equivalent to

$$(9') \qquad \hat{T}_t + \hat{T}_{xxx} = 0, \qquad \hat{T}_{xx} + 2\hat{a}\hat{T}_x = 0, \qquad (\hat{T}_x + \hat{a}\hat{T} - \hat{T}\hat{a})(I - \hat{P}) = 0,$$

$(\hat{N} = \hat{a})$, and equalities

$$(10) \qquad \hat{\gamma} = \hat{\Gamma}^{-1}\hat{\Gamma}_x = (I + \hat{T})^{-1}[a(I + \hat{T}) + \hat{T}_x]$$

are valid for the logarithmic derivative $\hat{\gamma}$. Therefore, the main statement of this subsection can be worded in the following equivalent form.

Let an operator function $\hat{T} = \hat{T}(x,t)$ in a certain domain of the x,t-plane satisfy Eqs. (9′), and let operators $I + \hat{T}(x,t)$ be invertible. Then the right-hand side of Eq. (10) is a solution of the operator Eq. (4), while its one-dimension projection $u(x,t)$ satisfies the scalar Eq. (1),

$$\left(\widehat{P}(I+\widehat{T})^{-1}\left[a(I+\widehat{T})+\widehat{T}_x\right]\widehat{P}=u(x,t)\widehat{P}\right).$$

5. In order to implement the above general scheme, one must select an arbitrary algebra of operator valued functions and find compatible solutions of Eq. (9) belonging to this algebra. The operator functions (6) always satisfy the first two equations. They evidently satisfy the third equation as well, for $\widehat{N}=\hat{a}$, if for the operator coefficient $\hat{m}$ one takes a solution of the equation $(\hat{a}\hat{m}+\hat{m}\hat{a})(I-\widehat{P})=0$, which is clearly equivalent to

$$\hat{a}\hat{m}+\hat{m}\hat{a}=\hat{r},\widehat{P}, \tag{11}$$

where $\hat{r}$ is an arbitrary constant operator.

Thus, seeking compatible solutions of Eq. (9) is reduced to choosing constant operators $\hat{a}$, $\hat{m}$, $\hat{r}$, $\widehat{P}$ that satisfy Eq. (11). Let the operators $\hat{a}$ and $\widehat{P}$ be arbitrary, while the operator $\hat{m}_0$ is found from the equation $\hat{a}\hat{m}_0+\hat{m}_0\hat{a}=\widehat{P}$. Then the operators $\hat{m}=\hat{r}\hat{m}_0$ shall clearly satisfy Eq. (11), provided that the operator $\hat{r}$ commutes with $\hat{a}$. So for the fixed operators $\hat{a}$ and $\widehat{P}$, one obtains solutions of Eq. (1) depending on the operator parameter $\hat{r}$ commutative with $\hat{a}$; that is, roughly speaking, the solutions depend on an arbitrary function defined on the spectrum of the operator $\hat{a}$. The broader the spectrum of this operator, the richer is the set of solutions of Eq. (1).

Now we give an example showing how the method is implemented.

As a linear space in which operators of an auxiliary algebra will act, we take a countable normed space of real infinitely differentiable functions $f(\xi)$, all of whose derivatives vanish at $\xi\to\infty$ faster than any negative power of ξ. Since constants belong to this space, the operator $\widehat{P}$ can be taken as a projection operator onto constant functions defined by the equality $\widehat{P}(f(\xi))=f(0)$. As the operator $\hat{a}=\widehat{N}$, we take an operator of differentiation with respect to ξ:

$$\hat{a}(f(\xi))=\widehat{N}(f(\xi))=-\frac{\partial}{\partial\xi}f(\xi)$$

(it will be convenient to set the minus sign).

We shall seek the operator function $\widehat{T}=\widehat{T}(x,t)$ satisfying Eq. (9′) in the form of the integral operator with the kernel $T(\xi,\eta;x,t)$:

$$\widehat{T}(f(\xi))=\int_0^\infty T(\xi,\eta;x,t)f(\eta)d\eta. \tag{12}$$

From the equation $\widehat{T}_{xx}+2\hat{a}\widehat{T}_x=0$ $(\hat{a}=-\frac{\partial}{\partial\xi})$, it follows that the kernel $T(\xi,\eta;x,t)$ must satisfy the partial differential equation

$$\frac{\partial}{\partial x}\left(\frac{\partial}{\partial x}T(\xi,\eta;x,t)-2\frac{\partial}{\partial\xi}T(\xi,\eta;x,t)\right)=0,$$

whose general solution has the form

$$T(\xi,\eta;x,t) = R(2x+\xi,\eta;t) + S(\xi,\eta;t).$$

Since the domain of values of the operator $(I-\widehat{P})$ consists of infinitely differential functions $f(\xi)$ that satisfy the boundary condition $f(0)=0$, the equation $(\widehat{T}_x + \hat{a}\widehat{T} - \widehat{T}a)(I-\widehat{P}) = 0$ means that the identity

$$\int_0^\infty \{R_\xi(2x+\xi,\eta;t) - S_\xi(\xi,\eta;t)\} f(\eta)d\eta + \int_0^\infty \{R(2x+\xi,\eta;t) + S(\xi,\eta;t)\} f'(\eta)d\eta = 0$$

must hold for these functions, whence, on integrating by parts, we obtain

$$\int_0^\infty \{R_\xi(2x+\xi,\eta;t) - R_\eta(2x+\xi,\eta;t) - S_\xi(\xi,\eta;t) - S_\eta(\xi,\eta;t)\} f(\eta)d\eta = 0.$$

Thus, the kernels R and S must satisfy the equations $R_\xi - R_\eta = 0$, $S_\xi + S_\eta = 0$. Having solved the equations, we find the following expressions for the kernels: $R = R(z;t), S = S(z_1;t)$, where $z = 2x+\xi+\eta$, $z_1 = \xi - \eta$. Hence,

$$T = R(2x+\xi+\eta;t) + S(\xi-\eta;t).$$

To let the operator with such a kernel satisfy the equation $\widehat{T}_t + \widehat{T}_{xxx} = 0$, we demand that the function S be independent of t and that the function $R(z;t)$ satisfy the equation $R_t + 8R_{zzz} = 0$.

Besides, the equalities

$$\lim_{z\to+\infty} z^p \frac{\partial^k}{\partial z^k} R(z;t) = \lim_{z_1\to+\infty} z^p \frac{\partial^k}{\partial z_1^k} S(z_1) = 0$$

must hold for all values of $p, k = 0,1,2,\ldots$, together with the inequality

$$\int_{-\infty}^\infty |S(\tau)|d\tau < \infty$$

which will guarantee correctness of the definition of the operator $\widehat{T}$ (see Eq. (12)).

Further, Eq. (10) implies that the function $\gamma(\xi) = \gamma(\xi,x;t) = \hat{\gamma}(1)$ can be found from the equation

$$\gamma(\xi) + \int_0^\infty [R(2x+\xi+\nu;t) + S(\xi-\nu)] f(\nu)d\nu + S(\xi) + R(2x+\xi;t) = 0 \qquad (0 \leqslant \xi < \infty),$$

and the one-dimensional projection $u(x,t)$ of the operator $\hat{\gamma}$ equals $\gamma(0;x,t)$.

In particular, when $S \equiv 0$, and

$$R(z;t) = \int_0^\infty e^{-\lambda(z-8\lambda^2 t)} d\mu(\lambda) + \frac{1}{2\pi}\int_{-\infty}^{\infty} r(\lambda)e^{i\lambda(z+8\lambda^2 t)} d\lambda,$$

$(r(\lambda) = \overline{r(-\lambda)}$, we obtain the standard equation of the inverse problem of scattering theory. (Invertibility of the operator $I+\widehat{T}$ is guaranteed by nonnegativity of the measure $d\mu$ and by the inequality $|r(\lambda)| \leqslant 1$). Solutions to the KdV equation obtained in such a fashion tend to zero when $x \to +\infty$.

It is easy to generalize this approach by taking the entire real axis instead of the semi-axis $0 \leqslant \xi < \infty$. In this case, the operators $\widehat{T}$ acquire the form

$$\widehat{T}(f(\xi)) = \int_0^\infty [R_1(2x+\xi+\eta;t) + S_1(\xi-\eta)]f(\eta)d\eta$$
$$+ \int_{-\infty}^0 [R_2(2x+\xi+\eta;t) + S_2(\xi-\eta)]f(\eta)d\eta.$$

We have not considered properties of these operators and resulting solutions of the KdV equation.

It is clear that other realizations of the above general scheme are possible. In Chapters 2 and 3, we consider realizations in finite-dimensional and separable Hilbert spaces, respectively. Note that different realizations can lead to the same solutions of nonlinear equations. It would be interesting to classify possible realizations from this viewpoint.

In Chapter 4 we investigate certain classes of solutions of nonlinear equations obtained in our realization of the general scheme of Hilbert spaces. The classes contain not only known solutions found by methods of the inverse problem, of the Riemann problem, or by methods of algebraic geometry, but other solutions as well, irreducible to the above-mentioned ones.

Of course, the suggested method is applicable not exclusively to the KdV equation, but to other nonlinear problems as well. The important requirements of the method are algebraic properties of differentiation operations and related properties of logarithmic derivatives $\widehat{\Gamma}^{-1}\partial\widehat{\Gamma}$. For this reason, we at first consider nonlinear equations in the abstract ring with differentiation, which permits to separate a purely algebraic treatment. This is done in Chapter 1.

Note, finally, that basic technical difficulties arise in solving Eq. (11) and in providing for invertibility of operators $\widehat{\Gamma}(\text{or}(I+\widehat{T}))$.

CHAPTER 1

THE GENERAL SCHEME

§1 Generalized Derivation and Logarithmic Derivatives

An associative ring is a set K for whose elements operations of addition $x+y$ and multiplication xy are defined that satisfy the conditions:

1. with respect to addition: K is an Abelian group;
2. multiplication is associative: $(xy)z = x(yz)$;
3. addition and multiplication are related by the distributive law:

$$z(x+y) = zx + zy, \qquad (x+y)z = xz + yz.$$

Such an element e that satisfies the equation $x = ex = xe$, for all $x \in K$, is called the unity of the ring. A ring cannot possess more than one unity, and any ring can be extended by adjoining the unity. Henceforth, we shall denote a ring with a unity by *ring*.

An element $x \in K$ is called invertible if there exists an element $x^{-1} \in K$, such that $x^{-1}x = xx^{-1} = e$. In this case, x^{-1} is said to be the inverse element of x; x^{-1} is determined by x in the unique way. The set of all inverse elements of the ring K is denoted by K^{-1}. The set of all elements of the ring that are commutative with all its elements is called the *centre of the ring* and is denoted by $Z(K)$.

A mapping L of the ring K into itself is called an *operator* if $L(x+y) = L(x)+L(y)$ for all $x, y \in K$. The set of all operators is denoted by $\mathcal{L}(K)$; it forms a ring with respect to the conventional operations of addition and multiplication for operators. The operator I is defined by the equality $I(x) = x$; it plays the role of the unity of the ring $\mathcal{L}(k)$.

An operator $\alpha \in \mathcal{L}(K)$ is called an *automorphism* if it performs a one-to-one mapping of the ring K onto itself and if $\alpha(xy) = \alpha(x)\alpha(y)$ for all $x, y \in K$. The set of all automorphisms is denoted by $\mathrm{Aut}(K)$ and forms a group with respect to multiplication. All automorphisms transfer the unity of the ring into itself.

DEFINITION 1.1.1: An operator $\partial \in \mathcal{L}(K)$ is called a generalized derivation (g.d.) in the ring K if there exists an automorphism $\alpha \in \mathrm{Aut}(K)$, such that for all $x, y \in K$,

$$\partial(x, y) = \partial(x)\alpha(y) + x\partial(y).$$

Setting $x = y = e$, we find that $\partial(e) = 0$ for any g.d. The set of all g.d.'s with the given automorphism α is denoted by $\mathrm{Der}(\alpha)$ and forms an Abelian group with respect to the addition of operators. In particular, $\mathrm{Der}(I)$ is a set of usual derivations that satisfy the identity

$$\partial(x, y) = \partial(x)y + x\partial(y).$$

Here are examples of operators, automorphisms and g.d.'s. Every element $a \in K$ generates the right, a_r, and the left, a_l, multiplication operators defined by the equalities

$$a_r(x) = xa, \qquad a_l(x) = ax.$$

Associativity of the ring implies that the operators a_r and a_l are commutative for any $a, b \in K : a_r b_l = b_l a_r$. It is obvious that $e_r = e_l = I$ and, if a is invertible, $(a^{-1})_r = (a_r)^{-1}, (a^{-1})_l = (a_l)^{-1}$. The operator $a_l - a_r$ is called a commutator with the element a and is denoted by $[a, \cdot]$:

$$[a, x] = (a_l - a_r)(x) = ax - xa.$$

Since

$$[a, xy] = axy - xya = (ax - xa)y + x(ay - ya) = [a, x]y + x[a, y],$$

then $[a, \cdot] \in \mathrm{Der}(I)$. This operator is called an inner derivation in the ring K.

Every invertible element $a \in K$ generates an inner automorphism $\alpha(x) = axa^{-1} = a_l a_r^{-1}(x)$ in the ring K.

An example of the g.d. from $\mathrm{Der}(\beta)(\beta \in \mathrm{Aut}(K))$ is the operator $\partial_\beta = \beta - I$, as

$$(\beta - I)(xy) = \beta(x)\beta(y) - xy = (\beta - I)(x)\beta(y) + x(\beta - I)(y).$$

If $\beta = b_l b_r^{-1}$, then the operator $\partial_\beta = \beta - I$ is called an inner g.d. in the ring K. Note also that if $\partial \in \mathrm{Der}(\alpha)$ and $b \in K^{-1}$, then the operator $b_r^{-1}\partial$ is a g.d. from $\mathrm{Der}(\beta)$, with the automorphism β defined by the equality $\beta(x) = b\alpha(x)b^{-1}$, i.e., $\beta = b_l b_r^{-1}\alpha$.

Let $\mathrm{Mat}_N(C^\infty(M \times G))$ be the set of all $N \times N$ square matrices $A = (a_{ik})(i, k = 1, \dots N)$ whose elements $a_{ik} = a_{ik}(m; x_1, \dots, x_p)$ are infinitely differentiable functions with respect to the variables $x_1, \dots, x_p$ that are defined on the direct product $M \times G$ of an arbitrary set M by a domain $G \subset R^p$ of the p-dimensional Euclidian space $(m \in M, (x_1, \dots, x_p) \in G)$. By defining the conventional matrix addition and multiplication in this space, we shall obtain a typical ring possessing, besides inner g.d.'s and automorphisms, also outer ones defined by the equalities

$$\partial_j(A) = \left(\frac{\partial a_{ik}}{\partial x_j}\right), \qquad \beta(A) = (a_{ik}(\beta(m); x_1, \dots, x_p))$$

where β is an arbitrary one-to-one mapping of the set M onto itself.

DEFINITION 1.1.2: Let Γ be an invertible element of K; the element

$$\gamma = \Gamma^{-1}\partial(\Gamma)$$

is called the logarithmic derivative of Γ with respect to the g.d. $\partial \in \mathrm{Der}(\alpha)$.

Invertible elements $\Gamma \in K$ generate also inner automorphisms in the ring $\mathcal{L}(K)$ that transfer operators $L \in \mathcal{L}(K)$ into operators

$$\tilde{L} = \Gamma_l^{-1} L \Gamma_l.$$

Obviously, the above operators satisfy the equality

$$\tilde{L}(e) = \Gamma^{-1} L(\Gamma).$$

Let $\partial \in \mathrm{Der}(\alpha)$, $\Gamma \in K^{-1}$, and $\tilde{\partial} = \Gamma_l^{-1}\partial\Gamma_l$. Then,

$$\begin{aligned}\tilde{\partial}(x) &= \Gamma^{-1}\partial(\Gamma x) = \Gamma^{-1}\{\partial(\Gamma)\alpha(x) + \Gamma\partial(x)\}\\ &= \gamma\alpha(x) + \partial(x) = (\gamma_l\alpha + \partial)(x).\end{aligned}$$

Thus, the operator $\tilde{\partial} = \Gamma_l^{-1}\partial\Gamma_l$ is expressed in terms of the logarithmic derivative $\gamma = \Gamma^{-1}\partial(\Gamma)$ in the following way,

$$\tilde{\partial} = \Gamma_l^{-1}\partial\Gamma_l = \gamma_l\alpha + \partial. \tag{1.1.1}$$

This is one of the basic formulas.

Differential operators in the ring K are the operators $D \in \mathcal{L}(K)$ that can be represented in the form,

$$D = b_r^0 + b_r^{p_1}\partial_{p_1} + b_r^{p_1,p_2}\partial_{p_1}\partial_{p_2} + \cdots + b_r^{p_1,\ldots,p_N}\partial_{p_1,\ldots,\partial_{p_N}}, \tag{1.1.2}$$

where

$$\partial_{p_j} \in \mathrm{Der}(\alpha_{p_j});\, b^0, b^{p_1}, \ldots, b^{p_1,\ldots,p_N} \in K,$$

the indices p_j take on all the values from 1 to n independently, and summation is carried over the repeated indices. The set of such operators is denoted by $D(\partial_1, \ldots, \partial_n)$ and is a subring of the ring $\mathcal{L}(K)$; the set forms a minimal subring containing all left multiplication operators and the given g.d.'s $\partial_1, \partial_2, \ldots, \partial_n$.

Equations of the form $D(x) = 0$, where $D \in D(\partial_1, \ldots, \partial_n)$, are called linear differential equations in the ring K.

Since the operators Γ_l^{-1}, Γ_l commute with $b_r^{p_1,\ldots,p_j}$, it follows that the operator $\tilde{D} = \Gamma_l^{-1} D \Gamma_l$ has the form

$$\tilde{D} = b_r^0 + b_r^{p_1}\tilde{\partial}_{p_1} + \cdots + b_r^{p_1,\ldots,p_N}\tilde{\partial}_{p_1}\tilde{\partial}_{p_2}\ldots\tilde{\partial}_{p_N}, \tag{1.1.3}$$

where

$$\tilde{\partial}_{p_j} = (\gamma_{p_j})_l\alpha_{p_j} + \partial_{p_j}, \qquad \gamma_{p_j} = \Gamma^{-1}\partial_{p_j}(\Gamma).$$

Since

$$\partial_{p_j}(e) = \gamma_{p_j}\alpha_{p_j}(e) + \partial_{p_j}(e)\gamma_{p_j},$$

and the indices p_j take on the values from 1 to n, the last formula implies that $\tilde{D}(e)$ is the function of logarithmic derivatives

$$\gamma_m = \Gamma^{-1}\partial_m(\Gamma), \qquad 1 \leqslant m \leqslant n,$$

and

$$\widetilde{D}(e) = \widetilde{D}(\gamma_1, \gamma_2, \ldots, \gamma_n),$$

respectively. This fact and the equality $\Gamma^{-1}D(\Gamma) = \widetilde{D}(e)$ yields the first of the two basic propositions of the method to be developed for the solution of nonlinear equations.

1. *If an invertible element Γ satisfies simultaneously several linear differential equations $D_p(\Gamma) = 0$, then its logarithmic derivatives satisfy the respective (generally, nonlinear) system of equations*

$$\widetilde{D}_p(e) = \widetilde{D}_p(\gamma_1, \gamma_2, \ldots, \gamma_n) = 0 \qquad (1 \leqslant p \leqslant m).$$

In what follows, we shall often make use of some particular cases of this general result.

Let

$$D = \partial_1\partial_2 - \partial_2\partial_1, \qquad \partial_i \in \mathrm{Der}(\alpha_i),$$
$$\gamma_i = \Gamma^{-1}\partial_i(\Gamma) \qquad (i = 1, 2), \qquad \Gamma \in K^{-1}.$$

Then, according to Eqs. (1.1.1) – (1.1.3),

$$\Gamma^{-1}D(\Gamma) = \widetilde{D}(e) = \tilde{\partial}_1\tilde{\partial}_2(e) - \tilde{\partial}_2\tilde{\partial}_1(e)$$
$$= \gamma_1\alpha_1(\gamma_2) + \partial_1(\gamma_2) - \gamma_2\alpha_2(\gamma_1) - \partial_2(\gamma_1)$$

and, if $D(\Gamma) = 0$, then the logarithmic derivatives of Γ satisfy the equation

$$\gamma_1\alpha_1(\gamma_2) + \partial_1(\gamma_2) - \gamma_2\alpha_2(\gamma_1) - \partial_2(\gamma_1) = 0.$$

In particular, if the g.d.'s ∂_1, ∂_2 are commutative, then $D(\Gamma) \equiv 0$, and the equation becomes an identity which is satisfied by logarithmic derivatives of any invertible element with respect to the commutative g.d.'s. For example, if $\partial_i \in \mathrm{Der}(I)$ $(i = 1, 2)$ and $\partial_1\partial_2 = \partial_2\partial_1$, then

$$\gamma_1\gamma_2 + \partial_1(\gamma_2) = \gamma_2\gamma_1 + \partial_2(x_1).$$

But if

$$\partial_1 \in \mathrm{Der}(I), \qquad \partial_2 = \partial_\alpha = \alpha - I \qquad (\alpha \in \mathrm{Aut}(K))$$

and $\partial_1\partial_\alpha = \partial_\alpha\partial_1$, then

$$\gamma_1\gamma_\alpha + \partial_1(\gamma_\alpha) = \gamma_\alpha\alpha(\gamma_1) + \partial_\alpha(\gamma_1),$$

where

$$\gamma_1 = \Gamma^{-1}\partial_1(\Gamma), \qquad \gamma_\alpha = \Gamma^{-1}\partial_\alpha(\Gamma).$$

Since $\partial_\alpha(\gamma_1) = \alpha(\gamma_1) - \gamma_1$, this identity is equivalent to

$$\partial_1(\gamma_\alpha) = (e + \gamma_\alpha)\alpha(\gamma_1) - \gamma_1(e + \gamma_\alpha).$$

Let the invertible element Γ satisfy the equation $D_1(\Gamma) = AD_2(\Gamma)$, where $D_1 D_2 \in D(\partial_1, \partial_2, \ldots, \partial_n)$ and A is an arbitrary element of the ring K. According to the previous reasoning, this equation is equivalent to

$$\Gamma \widetilde{D}_1(e) = A\Gamma \widetilde{D}_2(e).$$

If $\partial \in \mathrm{Der}(\alpha)$ and $\partial A = 0$, then, applying the operator ∂ to both sides of this equality, we find

$$\partial(\Gamma)\alpha(\widetilde{D}_1(e)) + \Gamma\partial(\widetilde{D}_2(e)) = A\partial(\Gamma)\alpha(\widetilde{D}_2(e)) + A\Gamma\partial(\widetilde{D}_2(e))$$

or

$$\gamma\alpha(\widetilde{D}_1(e)) + \partial(\widetilde{D}_1(e)) = \Gamma^{-1}A\Gamma\{\gamma\alpha(\widetilde{D}_2(e)) + \partial(\widetilde{D}_2(e))\}$$

where $\gamma = \Gamma^{-1}\partial(\Gamma)$. From both equalities we may exclude $\Gamma^{-1}A\Gamma$, provided that the element $D_2(\Gamma)$ and, therefore, the element $\widetilde{D}_2(e) = \Gamma^{-1}D_2(\Gamma)$, is invertible. Indeed, in this case, $\Gamma^{-1}A\Gamma = \widetilde{D}_1(e)(\widetilde{D}_2(e))^{-1}$, and

$$\gamma\alpha(\widetilde{D}_1(e)) + \partial(\widetilde{D}_1(e)) = \widetilde{D}_1(e)(\widetilde{D}_1(e))^{-1}\{\gamma\alpha(\widetilde{D}_2(e)) + \partial(\widetilde{D}_2(e))\},$$

that is,

$$\tilde{\partial}\widetilde{D}_1(e) = \widetilde{D}_1(e)(\widetilde{D}_2(e))^{-1}\tilde{\partial}(\widetilde{D}_2(e)).$$

In particular, if $D_2 = I$, then this equality is reduced to

$$\tilde{\partial}(\widetilde{D}_1(e)) = \widetilde{D}_1(e)\gamma,$$

and, if

$$D_2 = \partial_\beta + I, \qquad \partial_\beta = \beta - I, \qquad \beta \in \mathrm{Aut}(K)$$

and $\partial\partial_\beta = \partial_\beta\partial$, then it takes the form

$$\tilde{\partial}(\widetilde{D}_1(e)) = \widetilde{D}_1(e)\beta(\gamma).$$

Concluding this section, we shall give a table presenting the most frequently used formulas. The notations are as follows:

Γ—an arbitrary invertible element of the ring;
$\gamma, \gamma_1, \ldots, \gamma_\alpha$—its logarithmic derivative with respect to the g.d.'s $\partial, \partial_1, \ldots, \partial_\alpha$;
A—an arbitrary element of the ring;
D—an arbitrary differential operator in the ring;
$[x, y] = xy - yx$—a commutator, and
$\{x, y\} = xy + yx$—an anticommutator.

The left column of the table describes the conditions when the corresponding formulas on the right are valid.

This Table I is given in the Appendix.

§2 Examples of Nonlinear Equations

Throughout the sequel, Γ is assumed to be invertible and all considered g.d.'s are assumed to be commutative, from which follows the permutability of the similar operators

$$\tilde{\partial}_0 = \Gamma_l^{-1}\partial_0\Gamma_l, \qquad \tilde{\partial} = \Gamma_l^{-1}\partial\Gamma_l, \ldots, \tilde{\partial}_\alpha = \Gamma_l^{-1}\partial_\alpha\Gamma_l.$$

References to the formulas of Table I are given as I(1), I(2), and so forth. Finally, in deriving nonlinear equations solved by logarithmic derivatives, we shall now and then introduce auxiliary elements ($u = \partial\gamma$ or $u = [B, \gamma], v = \{B, \gamma\}$, etc.) which will allow us to write the final result in a more compact and conventional form.

1 *Korteweg–de Vries (KdV) Equation and its Generalizations*

Let Γ satisfy the equation

$$(\partial_0 + \partial^3)\Gamma = C\Gamma, \tag{1.2.1}$$

where $\partial_0, \partial \in \mathrm{Der}(I)$ and $\partial C = 0$. Then formula I(3) implies that

$$(\tilde{\partial}_0(e) + \tilde{\partial}^3(e))\gamma = \tilde{\partial}(\tilde{\partial}_0(e) + \tilde{\partial}^3(e)),$$

and since $\tilde{\partial}\tilde{\partial}_0 = \tilde{\partial}_0\tilde{\partial}$, we have

$$\tilde{\partial}_0\tilde{\partial}(e) - \tilde{\partial}_0(e)\gamma + \tilde{\partial}^4(e) - \tilde{\partial}^3(e)\gamma = 0.$$

Since, according to I(1),

$$\tilde{\partial}_0(e) = \gamma_0, \qquad \tilde{\partial}(e) = \gamma$$

and

$$\tilde{\partial}_0\tilde{\partial}(e) - \tilde{\partial}_0(e)\gamma = \gamma_0\gamma + \partial_0(\gamma) - \gamma_0\gamma = \partial_0\gamma,$$

and, according to I(2),

$$\tilde{\partial}^4(e) - \tilde{\partial}^3(e)\gamma = 3(\gamma^2\partial\gamma + \gamma\partial^2\gamma + \partial\gamma\partial\gamma) + \partial^3\gamma,$$

this equation is equivalent to

$$\partial_0\gamma + \partial^3\gamma + 3(\gamma^2\partial\gamma + \gamma\partial^2\gamma + \partial\gamma\partial\gamma) = 0. \tag{1.2.1'}$$

Thus, if an element Γ satisfies Eq. (1.2.1), then its logarithmic derivative satisfies Eq. (1.2.1').

By literally repeating the derivation of Eq. (1.2.1'), the reader can convince himself that if an element Γ satisfies the equation

$$(\partial_1 + \partial^2)\Gamma = A\Gamma, \tag{1.2.2}$$

where $\partial_1, \partial \in \mathrm{Der}(I), \partial(A) = 0$, then its logarithmic derivative $\gamma = \Gamma^{-1}\partial\Gamma$ satisfies the equation

$$\partial_1\gamma + \partial^2\gamma + 2\gamma\partial\gamma = 0. \tag{1.2.2$'$}$$

So, if Γ satisfies both Eq. (1.2.1) and Eq. (1.2.2) simultaneously, then its logarithmic derivative $\gamma = \Gamma^{-1}\partial\Gamma$ satisfies both Eq. (1.2.1′) and Eq. (1.2.2′). Therefore, we can eliminate all terms containing γ from Eq. (1.2.1′), retaining only those that contain nothing but derivatives of γ. To do this, we apply the operator ∂ to Eq. (1.2.2′)

$$\partial\partial_1\gamma + \partial^3\gamma + 2\gamma\partial^2\gamma + 2\partial\gamma\partial\gamma = 0.$$

Multiplying Eq. (1.2.2') by 2γ from the left, we have

$$2\gamma\partial_1\gamma + 2\gamma\partial^2\gamma + 4\gamma^2\partial\gamma = 0.$$

Summing these equalities,

$$4(\gamma^2\partial\gamma + \gamma\partial^2\gamma + \partial\gamma\partial\gamma) + \partial^3\gamma - 2\partial\gamma\partial\gamma + \partial\partial_1\gamma + 2\gamma\partial_1\gamma = 0$$

and comparing the result with Eq. (1.2.1′), we find

$$4\partial_0\gamma + \partial^3\gamma + 6\partial\gamma\partial\gamma - 3(\partial\partial_1\gamma + 2\gamma\partial_1\gamma) = 0.$$

Applying the operators ∂ and $3\partial_1$ to the latter equation and to Eq. (1.2.2′), respectively, and summing the results, we arrive at the equation

$$\partial(4\partial_0\gamma + \partial^3\gamma + 6\partial\gamma\partial\gamma) + 3\partial_1^2\gamma + 6[\partial_1\gamma, \partial\gamma] = 0, \tag{1.2.3}$$

which is satisfied by the logarithmic derivative $\gamma = \Gamma^{-1}\partial\Gamma$, if the element Γ solves Eqs. (1.2.1) and (1.2.2) simultaneously. Eq. (1.2.3) is called the Kadomtsev–Petriashvili (KP) equation.

If $\partial_1 = \lambda_l\partial$, where $\lambda \in Z(K)$ and $\partial\lambda = 0$, then the KP equation is reduced to the KdV equation

$$4\partial_0 u + \partial^3 u + 6(u\partial(u) + \partial(u)u) + 3\lambda^2\partial u = 0$$

in $u = \partial\gamma$.

If $\partial_0 = \lambda_l\partial$, then it is reduced to the nonlinear string equation

$$4\lambda\partial^2\gamma + \partial^4\gamma + 6(\partial\gamma\partial^2(\gamma) + \partial^2(\gamma)\partial\gamma) + 3\partial_1^2\gamma + 6[\partial_1\gamma, \partial\gamma] = 0.$$

Let us assume now that Γ simultaneously solves Eq. (1.2.1) and the equation

$$\partial\Gamma B = A\Gamma, \tag{1.2.4}$$

where $\partial A = \partial B = 0$ and $B^2 = e$. Then $\partial^2\Gamma B = A\partial\Gamma = A^2\Gamma B$ and $\partial^2\Gamma = A^2\Gamma$, whence it follows that Γ also satisfies the equation of the form (1.2.2), where one must set $\partial_1 = 0$ and substitute A for A^2. Whence, according to the aforesaid, it follows that $\gamma = \Gamma^{-1}\partial\Gamma$ satisfies the equation obtained from Eq. (1.2.3) with $\partial_1 = 0$, i.e.,

$$4\partial_0\gamma + \partial^3\gamma + 6\partial\gamma\partial\gamma = 0. \tag{1.2.4'}$$

Further, from Eq. (1.2.4) and formula I(3) it follows that $\tilde{\partial}(\tilde{\partial}(e)B) = \tilde{\partial}(e)B\gamma$, i.e., $(\gamma^2 + \partial\gamma)B = \gamma B\gamma$, or

$$\partial\gamma B = -\gamma[\gamma, B].$$

Setting $u = [\gamma, B]$ and commuting both sides of this equation with B, we find

$$\partial u B = -u^2 - 2\gamma[\gamma, B]B = -u^2 + 2\partial\gamma,$$

whence

$$2\partial\gamma = u^2 + \partial u B$$

since $\partial B = 0$ and, according to I(8), u and ∂u anticommutate with B. Finally, making use of the latter equality and commutating Eq. (1.2.4') with B, we obtain

$$\begin{aligned} 0 &= 4\partial_0 u + \partial^3 u + 3(2\partial\gamma\partial u + \partial u 2\partial\gamma) \\ &= 4\partial_0 u + \partial^3 u + 3\{u^2 + \partial u B)\partial u + \partial u(u^2 + (\partial)uB)\} \\ &= 4\partial_0 u + \partial^3 u + 3(u^2\partial u + \partial u u^2). \end{aligned}$$

Therefore, if Γ simultaneously solves Eqs. (1.2.1) and (1.2.4), then $u = [\gamma, B]$, where $\gamma = \Gamma^{-1}\partial\Gamma$ satisfies the modified KdV equation

$$4\partial_0 u + \partial^3 u + 3(u^2\partial u + \partial u u^2) = 0, \tag{1.2.5}$$

2 *Nonlinear Schrödinger Equation and Heisenberg Equation*

Let Γ satisfy the equations

$$\partial_0\Gamma + \partial^2\Gamma B = C\Gamma, \qquad \partial_1\Gamma + \partial\Gamma B = A\Gamma, \tag{1.2.6}$$

where $\partial_0, \partial, \partial_1 \in \mathrm{Der}(I), \partial A = \partial B = \partial C = \partial_0 B = \partial_1 B = 0, B^2 = e$. Then, according to I(3),

$$\begin{aligned} \tilde{\partial}(\tilde{\partial}_0(e) + \tilde{\partial}^2(e)B) - (\tilde{\partial}_0(e) + \tilde{\partial}^2(e)B)\gamma = 0, \\ \tilde{\partial}(\tilde{\partial}_1(e) + \tilde{\partial}(e)B) - (\tilde{\partial}_1(e) + \tilde{\partial}(e)B\gamma = 0, \end{aligned}$$

and, therefore,

$$\begin{aligned} \tilde{\partial}_0\tilde{\partial}(e) - \tilde{\partial}_0(e)\gamma + (\tilde{\partial}^3(e) - \tilde{\partial}^2(e)\gamma)B + \tilde{\partial}^2(e)[\gamma, B] = 0, \\ \tilde{\partial}_1\tilde{\partial}(e) - \tilde{\partial}_1(e)\gamma + (\tilde{\partial}^2(e) - \tilde{\partial}(e)\gamma)B + \tilde{\partial}(e)[\gamma, B] = 0 \end{aligned}$$

whence, using formula I(2), we arrive at the equations

$$\partial_0\gamma + (2\gamma\partial\gamma + \partial^2\gamma)B + (\gamma^2 + \partial\gamma)[\gamma, B] = 0 \tag{1.2.7}$$

$$\partial_1\gamma + \partial\gamma B + \gamma[\gamma, B] = 0 \tag{1.2.7'}$$

which are satisfied by $\gamma = \Gamma^{-1}\partial\Gamma$.

Applying the operator ∂_1 to Eq. (1.2.7′) and taking into account that $B^2 = e, B[\gamma, B] = -[\gamma, B]B$, we find

$$\begin{aligned} 0 &= \partial_1^2\gamma + \partial\partial_1\gamma B + \partial_1\gamma[\gamma, B] + \gamma[\partial_1\gamma, B] \\ &= \partial_1^2\gamma - \partial(\partial\gamma + \gamma[\gamma, B]B) - (\partial\gamma B + \gamma[\gamma, B])[\gamma, B] \\ &\quad - \gamma[\partial\gamma B + \gamma[\gamma, B], B] \\ &= (\partial_1^2 - \partial^2)\gamma - \partial\gamma[\gamma, B]B - \gamma[\partial\gamma, B]B \\ &\quad - \partial\gamma B[\gamma, B] - \gamma[\gamma, B]^2 - \gamma[\partial\gamma, B]B - \gamma[\gamma, B]^2 - 2\gamma^2[\gamma, B]B \\ &= (\partial_1^2 - \partial^2)\gamma - 2(\gamma[\partial\gamma, B]B + \gamma[\gamma, B]^2 + \gamma^2[\gamma, B]B). \end{aligned}$$

Therefore,

$$2\gamma^2[\gamma, B] = (\partial_1^2 - \partial^2)\gamma B - 2(\gamma[\partial\gamma, B] + \gamma[\gamma, B]^2 B),$$

which enables us to eliminate the term $\gamma^2[\gamma, B]$ from Eq. (1.2.7). On doing this, we find

$$\begin{aligned} 2\partial_0\gamma + (\partial_1^2 + \partial^2)\gamma B + 4\gamma\partial\gamma B + 2\partial\gamma[\gamma, B] \\ - 2\gamma[\partial\gamma, B] - 2\gamma[\gamma, B]^2 B = 0, \end{aligned}$$

that is

$$2\partial_0\gamma + (\partial_1^2 + \partial^2)\gamma B + 2\gamma\partial\{\gamma, B\} - 2(\partial\gamma B + \gamma[\gamma, B])[\gamma, B]B = 0.$$

Since $\partial\gamma B + \gamma[\gamma, B] = -\partial_1\gamma$ and

$$\partial\{\gamma, B\} = (-\partial_1\{\gamma, B\} + [\gamma, B]^2)B,$$

this expression is equivalent to

$$2\partial_0\gamma B + (\partial_1^2 + \partial^2)\gamma + 2\gamma[\gamma, B]^2 - 2\gamma\partial_1\{\gamma, B\} + 2\partial_1\gamma[\gamma, B] = 0,$$

whence, on commuting with B, we find

$$2\partial_0 u B + (\partial_1^2 + \partial^2)u + 2u^3 - 2\{u, \partial_1 v\} = 0,$$

where $u = [\gamma, B], v = \{\gamma, B\}$. Besides, Eq. (1.2.7′) implies

$$\partial_1 v + \partial v B = u^2, \qquad \partial_1 u + \partial u B = vu.$$

Thus, if Γ is a compatible solution of (1.2.6), then the elements

$$u = [\gamma, B], \qquad v = \{\gamma, B\}, \qquad (\gamma = \Gamma^{-1}\partial\Gamma)$$

satisfy the system of equations

$$\begin{aligned} 2\partial_0 uB + (\partial_1^2 + \partial^2)u + 2u^3 - 2\{u, \partial_1 v\} = 0, \\ \partial_1 v + \partial v B = u^2, \end{aligned} \tag{1.2.8}$$

which is an abstract form of the Davy–Stewardson system.

In particular, at $\partial_1 = 0$ this system is reduced to the nonlinear Schrödinger equation

$$2\partial_0 uB + \partial^2 u + 2u^3 = 0, \qquad u = [\gamma, B].$$

With $\partial_1 = 0$ and an invertible A, Eq. (1.2.6) yields one more nonlinear equation of interest. In this case, according to Eq. (1.2.7′),

$$\partial\gamma B = -\gamma[\gamma, B], \tag{1.2.9}$$

whence it follows that

$$\begin{aligned} (2\gamma\partial\gamma + \partial^2\gamma)B &= -2\gamma^2[\gamma, B] - \partial\gamma[\gamma, B] - \gamma[\partial\gamma, B] \\ &= -(\gamma^2 + \partial\gamma)[\gamma, B] - \gamma^2[\gamma, B] - \gamma[\partial\gamma, B] \\ &= -(\gamma^2 + \partial\gamma)[\gamma, B] + \gamma\partial\gamma B - \gamma\partial\gamma B + \gamma B\partial\gamma, \end{aligned}$$

that is

$$(2\gamma\partial\gamma + \partial^2\gamma)B + (\gamma^2 + \partial\gamma)[\gamma, B] = \gamma B\partial\gamma.$$

Substituting this expression into Eq. (1.2.6), we find

$$\partial_0\gamma + \gamma B\partial\gamma = 0. \tag{1.2.9′}$$

Further, it follows from Eq. (1.2.5) at $\partial_1 = 0$ that the element $\gamma = \Gamma^{-1}\partial\Gamma = \Gamma^{-1}A\Gamma B$ is invertible if $A \in K^{-1}$. Therefore Eqs. (1.2.9) and (1.2.9′) are equivalent to the following equations

$$\gamma^{-1}\partial\gamma = -[\gamma, B]B, \qquad \gamma^{-1}\partial_0\gamma = -B\partial\gamma,$$

from which we obtain the equalities

$$\begin{aligned} [\gamma^{-1}\partial\gamma, B] &= -2[\gamma, B], \\ [\gamma^{-1}\partial\gamma, [\gamma, B]] &= 2[\gamma, B]^2 B, \\ [\gamma^{-1}\partial_0\gamma, B] &= -B[\partial\gamma, B]. \end{aligned}$$

Using these equalities and setting $S = \gamma B\gamma^{-1}$, we find, in accordance with I(5),

$$\begin{aligned} \partial_0 S &= -\gamma B[\partial\gamma, B]\gamma^{-1}, \\ \partial S &= -2\gamma[\gamma, B]\gamma^{-1}, \\ \partial^2 S &= -2\gamma([\gamma^{-1}\partial\gamma, [\gamma, B]] + [\partial\gamma, B])\gamma^{-1} \\ &= -2\gamma(2[\gamma, B]^2 B + [\partial\gamma, B])\gamma^{-1}, \\ [S, \partial^2 S] &= -2\gamma\left[B, 2[\gamma, B]^2 B + [\partial\gamma, B]\right]\gamma^{-1} \\ &= -4\gamma B[\partial\gamma, B]\gamma^{-1}. \end{aligned}$$

Therefore, if Γ satisfies the system (1.2.6) with $\partial_1 = 0, A \in K^{-1}$, then the element

$$S = \gamma B \gamma^{-1} \qquad (\gamma = \Gamma^{-1}\partial\Gamma)$$

satisfies the abstract Heisenberg equation

$$4\partial_0 S = [S, \partial^2 S] \tag{1.2.10}$$

3 *Systems of Nonlinear Equations*

Let Γ satisfy the equations

$$\partial_1\Gamma = A_1\Gamma D_1, \qquad \partial_2\Gamma = A_2\Gamma D_2, \tag{1.2.11}$$

where $\partial_1, \partial_2 \in \mathrm{Der}(I)$, the elements A_1, A_2, D_1, D_2 are invertible, $\partial_1 D_2 = \partial_2 D_1 = \partial_1 A_2 = \partial_2 A_1 = 0, [D_1, D_2] = 0$ and

$$A_1 A_2 = \lambda_1 A_1 + \lambda_2 A_2 \qquad (\lambda_1, \lambda_2 \in Z(K)). \tag{1.2.11'}$$

Then the logarithmic derivatives $\gamma_i = \Gamma^{-1}\partial_i\Gamma = \Gamma^{-1}A_i\Gamma D_i (i = 1, 2)$ are invertible and, according to I(3),

$$\begin{aligned} \partial_j\gamma_i &= -\gamma_j\gamma_i + \gamma_i D_i^{-1}\gamma_j D_i \\ \partial_j(\gamma_i)\gamma_i^{-1} &= -\gamma_j + \gamma_i D_i^{-1}\gamma_j D_i\gamma_i^{-1} \end{aligned} \qquad (i \neq j) \tag{1.2.12}$$

Putting $u_i = \gamma_i D_i \gamma_i^{-1}$ and using that $\partial_j D_i = 0 (i \neq j)$, we find

$$\partial_j u_i = [\partial_j(\gamma_i)\gamma_i^{-1}, u_i]. \tag{1.2.13}$$

Further, it follows from Eq. (1.2.11) that for $i \neq j$,

$$\gamma_i D_i^{-1}\gamma_j D_j^{-1} = \lambda_i\gamma_i D_i^{-1} + \lambda_j\gamma_j D_j^{-1}$$

and, therefore,

$$\lambda_i D_j\gamma_j^{-1} + \lambda_j D_i\gamma_i^{-1} = e.$$

From here we deduce

$$\begin{aligned} \gamma_i D_i^{-1}\gamma_j D_i\gamma_i^{-1} &= \lambda_i\gamma_i D_j\gamma_i^{-1} + \lambda_j\gamma_j D_i\gamma_i^{-1} \\ &= \gamma_j + \lambda_i(\gamma_i D_j\gamma_i^{-1} - \gamma_j D_j\gamma_j^{-1}) \end{aligned}$$

whence, according to Eq. (1.2.9), we find

$$\partial_j(\gamma_i)\gamma_i^{-1} = \lambda_i(\gamma_i D_j\gamma_i^{-1} - \gamma_j D_j\gamma_j^{-1}).$$

Substituting these expressions into Eq. (1.2.13), we obtain

$$\begin{aligned} \partial_j u_i &= \lambda_i[\gamma_i D_j\gamma_i^{-1} - \gamma_j Dj\gamma_j^{-1}, \gamma_i D_i\gamma_i^{-1}] \\ &= \lambda_i\gamma_i[D_j, D_i]\gamma_i^{-1} - \lambda_i[\gamma_j D_j\gamma_j^{-1}, \gamma_i D_i\gamma_i^{-1}] \\ &= \lambda_i[u_i, u_j]. \end{aligned}$$

Thus, if Γ is a simultaneous solution of Eqs. (1.2.11) satisfying the condition (1.2.11′), then the elements

$$u_i = \gamma_i D_i \gamma_i^{-1} \qquad (\gamma_i = \Gamma^{-1}\partial_i \Gamma)$$

satisfy the system of equations

$$\partial_2 u_1 = \lambda_1 [u_1, u_2], \qquad \partial_1 u_2 = \lambda_2 [u_2, u_1] \tag{1.2.14}$$

which is known in chiral field theory.

Indeed, according to Eq. (1.2.11′),

$$e - 2\lambda_2 D_1 \gamma_1^{-1} = 2\lambda_1 D_2 \gamma_2^{-1} - e = v, \tag{1.2.14′}$$

and since $D_1\gamma_i^{-1} = \Gamma^{-1}A_i^{-1}\Gamma$, the element v admits such equivalent representation

$$v = \Gamma^{-1} C \Gamma, \qquad C = e - 2\lambda_2 A_1^{-1} = 2\lambda_1 A_2^{-1} - e.$$

Applying the operators ∂_1, ∂_2 to both sides of this equality, we find that $\partial_i v = [v, \gamma_i]$. Further, we deduce from Eq. (1.2.14′),

$$\begin{aligned} [v, \gamma_1] &= 2\lambda_2 (u_1 - D_1), \quad & [v, \gamma_2] &= 2\lambda_1 (D_2 - u_2), \\ v u_1 &= u_1 - D_1 + D_1 v, \quad & v u_2 &= -u_2 + D_2 + D_2 v, \end{aligned}$$

where $u_i = \gamma_i D_i \gamma_i^{-1}$. Hence, the element v satisfies the system $\partial_i v = 2\lambda_j (v u_i - D_i v)(i = 1, 2)$. If in the ring there exists an element $\mathcal{E}$ satisfying the equations $\partial_i \mathcal{E} = 2\lambda_j \mathcal{E} D_i$, then, setting $S = \mathcal{E} v$, we get

$$\partial_i S = \partial_i \mathcal{E} v + \mathcal{E} \partial_i v = \mathcal{E}(2\lambda_j D_i v + 2\lambda_j v u_i - 2\lambda_j D_i v) = 2\lambda_j S u_i.$$

Hence, if $S \in K^{-1}$, then

$$(\partial_1 S) S^{-1} (\partial_2 S) + (\partial_2 S) S^{-1} (\partial_1 S) = 4\lambda_1\lambda_2 S\{u_1, u_2\}$$

and, according to Eq. (1.2.14),

$$\partial_1\partial_2 S = 2\lambda_1 \partial_1 (S u_2 = 4\lambda_1\lambda_2 S u_1 u_2 - 2\lambda_1\lambda_2 S[u_1, u_2] = 2\lambda_1\lambda_2 S\{u_1, u_2\}.$$

Thus, the element $S = \mathcal{E}\Gamma^{-1}C\Gamma (C = e - 2\lambda_2 A_1^{-1}, \partial_i \mathcal{E} = 2\lambda_j \mathcal{E} D_i)$ satisfies the equation

$$2\partial_1\partial_2 S = (\partial_1 S) S^{-1} (\partial_2 S) + (\partial_2 S) S^{-1} (\partial_1 S)$$

which is an abstract form of the principal chiral field equation.

If $A_1 = A_2 = A, \partial_1 A = \partial_2 A = 0, \partial_1 D_2 = \partial_2 D_1 = 0$ in Eq. (1.2.11), and the elements D_1, D_2 are invertible, then

$$\gamma_1 D_1^{-1} = \gamma_2 D_2^{-1} = \Gamma^{-1} A \Gamma = u$$

and, according to I(4),

$$\partial_i u = -\gamma_i \Gamma^{-1} A \Gamma + \Gamma^{-1} A \Gamma \gamma_i = -u D_i u + u^2 D_i = u[u, D_i].$$

Therefore,

$$\partial_i[u, D_j] = [\partial_i u, D_j] = [u, D_j][u, D_i] + u[[u, D_i], D_j],$$

while permutability of D_1, D_2 implies the equality $[[u, D_i], D_j] = [[u, D_j]D_i]$. Hence, the elements

$$v_i = [u, D_i] = \gamma_i - D_i\gamma_i D_i^{-1} = [\gamma_i, D_i]D_i^{-1}$$

satisfy the system of equations

$$\partial_1 v_2 - \partial_2 v_1 = [v_2, v_1], \qquad [v_2, D_1] = [v_1, D_2], \tag{1.2.15}$$

which is equivalent to the equation of the N-wave problem:

$$\partial_1[u, D_2] - \partial_2[u, D_1] = [[u, D_2], [u, D_1]].$$

Now let Γ satisfy the equations

$$\partial_1\Gamma = (\partial\Gamma + A\Gamma)D_1, \qquad \partial_2\Gamma = (\partial\Gamma + A\Gamma)D_2,$$

where

$$\partial, \partial_1, \partial_2 \in \mathrm{Der}(I), \qquad D_1, D_2 \in K^{-1}, \qquad [D_1, D_2] = 0$$

and

$$\partial_1 D_2 = \partial_2 D_1 = \partial D_1 = \partial D_2 = \partial_1 A = \partial_2 A = 0.$$

Then $\gamma_1 D_1^{-1} = \gamma_2 D_2^{-1} = u$ and, according to I(3), I(6),

$$\begin{aligned}
0 &= \tilde{\partial}_j(\gamma_j D_i^{-1} - \gamma) - (\gamma_i D_i^{-1} - \gamma)\gamma_j \\
&= \gamma_j u + \partial_j u - \gamma_j\gamma - \partial_j\gamma - u\gamma_j + \gamma\gamma_j \\
&= uD_j u + \partial_j u - \partial\gamma_j - u^2 D_j \\
&= \partial_j u - \partial u D_j - u[u, D_j].
\end{aligned}$$

Commuting this equality with $D_i (i \neq j)$, we find

$$\partial_j[u, D_i] - \partial[u, D_i]D_j - [u, D_i][u, D_j] - u[[u, D_j], D_i] = 0$$

whence, using permutability of the elements D_1, D_2 and the equality that follows from it—$[[u, D_1], D_2] = [[u, D_2], D_1]$ —we deduce that the elements

$$v_i = [u, D_i] = \gamma_i - D_i\gamma_i D_i^{-1} = [\gamma_i, D_i]D_i^{-1}$$

satisfy the system of the equations

$$\begin{aligned}
\partial_2 v_1 - \partial_1 v_2 - \partial(v_1 D_2 - v_2 D_1) &= [v_1, v_2], \\
[v_1, D_2] &= [v_2, D_1]
\end{aligned} \tag{1.2.16}$$

which is equivalent to the equation

$$\partial_2[u, D_1] - \partial_1[u, D_2] - \partial([u, D_1]D_2 - [u, D_2]D_1) = [[u, D_1], [u, D_2]]$$

of the generalized N-wave problem.

4 *Nonlinear Equations Containing Automorphisms*

Let us denote by α an arbitrary automorphism of the ring K and by $\partial_\alpha = \alpha - I \in \mathrm{Der}(\alpha)$—the corresponding g.d.

Let Γ satisfy the equations

$$\partial\Gamma = (\partial_\alpha + I)\Gamma, \qquad \partial^2\Gamma + \Gamma = A(\partial_\alpha + I)\Gamma, \tag{1.2.17}$$

where $\partial \in \mathrm{Der}(I)$ and $\partial A = 0$. Then the logarithmic derivative $\gamma = \Gamma^{-1}\partial\Gamma = \Gamma^{-1}(\partial_\alpha + I)\Gamma = \Gamma^{-1}\alpha(\Gamma)$ is invertible: $\gamma^{-1} = \alpha(\Gamma^{-1})\Gamma$. Further, Eq. (1.2.17) and formula I(7) imply that γ satisfies the equations

$$\partial\gamma - \gamma\partial_\alpha(\gamma) = 0, \qquad \partial^2\gamma + 2\gamma\partial\gamma - (\gamma^2 + \partial\gamma + e)\partial_\alpha(\gamma) = 0,$$

which yield the equalities

$$\gamma^{-1}\partial\gamma = \partial_\alpha(\gamma), \qquad g^{-1}\partial^2\gamma - \gamma^{-1}\partial\gamma\gamma^{-1}\partial\gamma = \gamma^{-1}\partial_\alpha(\gamma) - \partial\gamma.$$

Hence

$$\begin{aligned} \partial(\gamma^{-1}\partial\gamma) &= \partial_\alpha\partial\gamma = \alpha(\partial\gamma) - \partial\gamma, \\ \partial(\gamma^{-1}\partial\gamma) &\equiv \gamma^{-1}\partial^2\gamma - \gamma^{-1}\partial\gamma\gamma^{-1}\partial\gamma \\ &= \gamma^{-1}\partial_\alpha(\gamma) - \partial\gamma \end{aligned} \tag{1.2.18}$$

and, therefore,

$$\begin{aligned} \alpha(\partial\gamma) &= \gamma^{-1}\partial_\alpha(\gamma) = \gamma^{-1}\alpha(\gamma) - e, \\ \partial\gamma &= \alpha^{-1}(\gamma^{-1})\gamma - e. \end{aligned}$$

Substituting these expressions into the right side of Eq. (1.2.18), we find that γ satisfies the equation

$$\partial(\gamma^{-1}\partial\gamma) = \gamma^{-1}\alpha(\gamma) - \alpha^{-1}(\gamma^{-1})\gamma, \tag{1.2.19}$$

which is the abstract form of the Toda lattice oscillation equation.

Now let Γ satisfy the equations

$$\partial_1(\partial\Gamma) = \Gamma, \qquad \partial\Gamma = A(\partial_\alpha + I)\Gamma,$$

where $\partial, \partial_1 \in \mathrm{Der}(I)$; the element A is invertible and $\partial A = \partial_1 A = 0$. In this case the logarithmic derivative $\gamma = \Gamma^{-1}\partial\Gamma$ is also invertible ($\gamma^{-1} = \alpha(\Gamma^{-1})A^{-1}\Gamma$), and the first of the equations implies the equalities

$$\gamma_1\gamma + \partial_1\gamma = \gamma\gamma_1 + \partial\gamma_1 = e,$$

while the second one and formula I(7) yield

$$\gamma_1\gamma + \partial_1\gamma - \gamma\gamma_1 - \gamma\partial_\alpha(\gamma_1) = 0, \qquad \gamma^2 + \partial\gamma - \gamma^2 - \gamma\partial_\alpha(\gamma) = 0.$$

So $e = \gamma(\gamma_1 + \partial_\alpha(\gamma_1)) = \gamma\alpha(\gamma_1), \partial\gamma = \gamma\partial_\alpha(\gamma)$ and, therefore,

$$\alpha(\gamma_1 = \gamma^{-1}, \qquad \gamma_1 = \alpha^{-1}(\gamma^{-1}), \qquad \gamma^{-1}\partial\gamma = \partial_\alpha(\gamma),$$

whence we find

$$\begin{aligned}\partial_1(\gamma^{-1}\partial\gamma) &= \partial_\alpha(\partial_1\gamma) = \partial_\alpha(e - \gamma_1\gamma) = -\partial_\alpha(\gamma_1\gamma)\\ &= -\alpha(\gamma_1)\alpha(\gamma) + \gamma_1\gamma = \alpha^{-1}(\gamma^{-1})\gamma - \gamma^{-1}\alpha(\gamma).\end{aligned}$$

Thus, γ satisfies the abstract analogue of the sine-Gordon equation

$$\partial_1(\gamma^{-1}\partial\gamma) = \alpha^{-1}(\gamma^{-1})\gamma - \gamma^{-1}\alpha(\gamma). \tag{1.2.20}$$

Finally, let Γ satisfy the equations

$$\partial\Gamma = (\partial_\alpha + I)^2\Gamma, \qquad \partial\Gamma + \Gamma = A(\partial_\alpha + I)\Gamma,$$

where $\partial \in \mathrm{Der}(I)$ and $\partial A = 0$. Since $(\partial_\alpha + I)^2 = \alpha^2 = \partial_{\alpha^2} + I$, the first equation is equivalent to $\partial\Gamma = (\partial_{\alpha^2} + I)\Gamma$, whence, making use of $I(7)$, we find that $\gamma^2 + \partial\gamma - \gamma^2 - \gamma\partial_{\alpha^2}(\gamma) = 0$, and so

$$\partial\gamma = \gamma\partial_{\alpha^2}(\gamma) = \gamma(\alpha^2(\gamma) - \gamma + \alpha(\gamma) - \alpha(\gamma)) = \gamma\alpha(\partial_\alpha\gamma) + \gamma\partial_\alpha\gamma.$$

The second equation and formula I(7) imply that $\gamma(\gamma + e) + \partial\gamma - (\gamma + e)\gamma - (\gamma + e)\partial_\alpha\gamma = 0$, that is

$$\partial\gamma = (\gamma + e)\partial_\alpha\gamma = \partial_\alpha\gamma + \gamma\partial_\alpha\gamma.$$

Hence, $\gamma = \Gamma^{-1}\partial\Gamma$ satisfy the equations

$$\partial\gamma = (\gamma + e)\partial_\alpha\gamma, \qquad \gamma\alpha(\partial_\alpha\gamma) = \partial_\alpha\gamma.$$

Having applied the operator ∂_α to the first of these equations, we find

$$\partial(\partial_\alpha\gamma) = \partial_\alpha(\gamma)\alpha(\partial_\alpha\gamma) + (\gamma + e)\partial_\alpha^2\gamma$$

and, since $\partial_\alpha^2\gamma = \alpha(\partial_\alpha\gamma) - \partial_\alpha\gamma$, then

$$\partial(\partial_\alpha\gamma) = (\partial_\alpha\gamma + e)\alpha(\partial_\alpha\gamma) - \partial_\alpha\gamma + \gamma\alpha(\partial_\alpha\gamma) - \gamma\partial_\alpha\gamma.$$

According to the second equation,

$$\begin{gathered}\gamma\alpha(\partial_\alpha\gamma) - (\partial_\alpha\gamma = 0,\\ \alpha^{-1}(\gamma)\partial_\alpha\gamma = \alpha^{-1}(\partial_\alpha\gamma) = \gamma - \alpha^{-1}(\gamma)\end{gathered}$$

and, therefore,

$$\gamma\partial_\alpha\gamma = \gamma^{-1}(\gamma)(\partial_\alpha\gamma + e)\partial_\alpha\gamma = \alpha^{-1}(\partial_\alpha\gamma)(\partial_\alpha\gamma + e)$$

and

$$\partial(\partial_\alpha\gamma) = (e + \partial_\alpha\gamma)\alpha(\partial_\alpha\gamma) - \alpha^{-1}(\partial_\alpha\gamma)(e + \partial_\alpha\gamma).$$

That is why the element $u = e + \partial_\alpha\gamma$ satisfies the equation

$$\partial u = u\alpha(u) - \alpha^{-1}(u)u, \tag{1.2.21}$$

which is a generalization of the Langmuir lattice oscillation equation. Note also that the equation $\partial\Gamma = (\partial_{\alpha^2} + I)\Gamma$ implies the invertibility of $\gamma = \Gamma^{-1}\partial\gamma = \Gamma^{-1}\alpha^{-2}(\Gamma)$ and the equality $\gamma\alpha(\partial_\alpha\gamma) = \partial_\alpha\gamma$ implies that $\gamma\alpha(u - e) = \alpha(\gamma) - \gamma$, that is, $\gamma\alpha(u) = \alpha(\gamma)$ and $u = \alpha^{-1}(\gamma^{-1})\gamma$, whence u can be expressed through γ in two different ways:

$$u = e + \partial_\alpha\gamma = \alpha^{-1}(\gamma^{-1})\gamma. \tag{1.2.21'}$$

5 *Equations of Higher Orders*

Let Γ satisfy the equation

$$\partial_0\Gamma + \sum_{k=1}^{N} 2^{k-1}\partial^k(\Gamma)C_k = 0, \qquad \partial^2\Gamma = A\Gamma, \tag{1.2.22}$$

where $\partial_0, \partial \in \mathrm{Der}(I), C_k \in Z(K), \partial_0 C_k = \partial C_k = \partial_0 A = \partial A = 0$. From the first equation and formula I(3) it follows that

$$\tilde{\partial}(\tilde{\partial}_0(e) + \sum_{k=1}^{N} \tilde{\partial}^k(e)2^{k-1}C_k) - (\tilde{\partial}_0(e) + \sum_{k=1}^{N} 2^{k-1}\tilde{\partial}^k(e)C_k)\gamma = 0,$$

and $\tilde{\partial}\tilde{\partial}_0(e) - \tilde{\partial}_0(e)\gamma = \gamma\gamma_0 + \partial\gamma_0 - \gamma_0\gamma = \partial_0\gamma$ implies that

$$\partial_0\gamma + \sum_{k=1}^{N} 2^{k-1}(\tilde{\partial}^{(k+1)}(e) - \tilde{\partial}^k(e)\gamma)C_k = 0. \tag{1.2.23}$$

From the second equation and equality (1.2.2') at $\partial_1 = 0$, it follows that

$$\partial^2\gamma + 2\gamma\partial\gamma = 0. \tag{1.2.24}$$

Employing this equality and formula I(2), we obtain

$$\tilde{\partial}(e) - \tilde{\partial}_0(e)\gamma = 0, \qquad \tilde{\partial}^2(e) - \tilde{\partial}(e)\gamma = \partial\gamma = \tilde{\partial}_0(e)\partial\gamma,$$
$$\tilde{\partial}^3(e) - \tilde{\partial}^2(e)\gamma = \partial^2\gamma + 2\gamma\partial\gamma = 0, \qquad \tilde{\partial}^4(e) - \tilde{\partial}^3(e)\gamma = \tilde{\partial}^2(e)\partial\gamma.$$

Thus, for $k = 0, 1$, the equalities

$$\widetilde{\partial}^{2k+1}(e) - \widetilde{\partial}^{2k}(e)\gamma = 0, \qquad \widetilde{\partial}^{2k+2}(e) - \widetilde{\partial}^{2k+1}(e)\gamma = \widetilde{\partial}^{2k}(e)\partial\gamma$$

hold, whence it follows by induction that they hold for all integer values of k, provided that γ satisfies Eq. (1.2.24).

Therefore, if Γ satisfies both equations (1.2.22), then Eq. (1.2.23) solved by logarithmic derivatives $\gamma = \Gamma^{-1}\partial\Gamma$ is equivalent to

$$\partial_0\gamma + \sum_{1\leqslant 2j+1\leqslant N} 2^{2j}\widetilde{\partial}^{2j}(e)\partial\gamma C_{2j+1} = 0. \tag{1.2.25}$$

We will show that Eq. (1.2.24) permits us to eliminate γ from the above equation, leaving only its derivatives $\partial_0\gamma, \partial\gamma, \partial^2\gamma, \ldots$. Indeed, having applied the operator ∂^n to both parts of the equality (1.2.24), we arrive at the formulas

$$2\gamma\partial^{n+1}\gamma = P_{n+1}(\partial\gamma, \partial^2\gamma, \ldots, \partial^{n+2}\gamma), \tag{1.2.25'}$$

where P_{n+1} are some polynomials in $\partial\gamma, \partial^2\gamma, \ldots, \partial^{n+2}\gamma$.

For instance,

$$\begin{aligned}
2\gamma\partial\gamma &= -\partial^2\gamma,\\
2\gamma\partial^2\gamma &= -2(\partial\gamma)^2 - \partial^3\gamma,\\
2\gamma\partial^3\gamma &= -4\partial\gamma\partial^2\gamma - 2\partial^2\gamma\partial\gamma - \partial^4\gamma,\\
2\gamma\partial^4\gamma &= -6\partial\gamma\partial^3\gamma - 6(\partial^2\gamma)^2 - 2\partial^3\gamma\partial\gamma - \partial^5\gamma.
\end{aligned}$$

Formulas (1.2.25) show that any expression $Q(\gamma, \partial\gamma, \partial^2, \gamma, \ldots)\partial\gamma$, where Q is a polynomial in $\gamma, \partial\gamma, \partial^2\gamma, \ldots$, can be rewritten in the form of another polynomial $R(\partial\gamma, \partial^2\gamma, \ldots)$ that depends only on derivatives of the element γ. In particular, since $\widetilde{\partial}^m(e) = (\gamma + \partial)\ldots(\gamma + \partial)\gamma$ is evidently a polynomial in $\gamma, \partial\gamma, \ldots$,

$$2^m\widetilde{\partial}^m(e)\partial\gamma = R_m(\partial\gamma, \partial^2\gamma, \ldots),$$

which permits us to rewrite Eq. (1.2.25) in such a form

$$\partial_0\gamma + \sum_{1\leqslant 2j+1\leqslant N} R_{2j}(\partial\gamma, \partial^2\gamma, \ldots)C_{2j+1} = 0. \tag{1.2.26}$$

For instance,

$$\begin{aligned}
R_2 &= 2^2\widetilde{\partial}^2(e)\partial\gamma = 6(\partial\gamma)^2 + \partial^3\gamma,\\
R_4 &= 2^4\widetilde{\partial}^4(e)\partial\gamma = 40(\partial\gamma)^3\\
&\qquad + 10(\partial\gamma\partial^3\gamma + (\partial^2\gamma)^2 + \partial^3\gamma\partial\gamma) + \partial^5\gamma,
\end{aligned}$$

and so forth.

Equations (1.2.26) are called highest KdV equations. We must bear in mind that all of them admit an additive group of transformations $\gamma \to \gamma + C$ consisting of elements C of the ring K that belong to kernels of the operators ∂_0, ∂, i.e., if γ satisfies an equation, then all elements $\gamma + C$ do the same.

§3 Projection Operation

In the previous sections we have demonstrated that logarithmic derivatives of solutions of some simultaneous linear differential equations satisfy the appropriate nonlinear equations. The solutions of nonlinear equations obtained in the ring K in such a manner are analogous to one-soliton solutions and depend on a small number of parameters belonging to this ring. One can, however, start from one-soliton solutions that belong to the ring K and, employing the projection operation, construct the solutions of the same nonlinear equation belonging to some subring $K_0 \subset K$; the greater the ring K compared to the subring K_0, the larger the set of solutions. The projection operation which is described in this section is the second of the two main components that comprise the method of solving nonlinear equations treated in this book.

We shall call constant those elements of the ring K which satisfy the equalities

$$\partial_j A = 0, \qquad \alpha_j(A) = A$$

for all g.d.'s $\partial_j \in \mathrm{Der}(\alpha_j)$ contained in the problem considered. Note that constant factors can be taken out of a g.d. sign:

$$\partial_j(xA) = \partial_j(x)A, \qquad \partial_j(Ax) = A\partial_j(x).$$

Let P denote a constant idempotent ($P^2 = P$) element of the ring K. The set of all elements of the form $PxP (x \in K)$ obviously makes a certain subring of the ring K that we shall denote by PKP. The mapping $x \longrightarrow PxP$ is called the operation of projection of the ring K onto its subring PKP, and the related operator $\widehat{P}(x) = PxP$ is called a projection operator. The idempotent P is the unity in the subring PKP, and all g.d.'s considered in the problem automatically remain g.d.'s in the subring PKP, while $\partial(PxP) = P\partial(x)P$.

Let

$$Q(y_1, y_2, \ldots) = \sum_{j_i} \partial_{k_1}^{m_1}(y_{j_1})\partial_{k_2}^{m_2}(y_{j_2}) \ldots A_{j_1, j_2, \ldots}$$

be an arbitrary polynomial in $y_i \in K$ with coefficients $A_{j_1, j_2, \ldots} \in K$. (By definition, $\partial^0(y) = y$).

Using projection operations for solving nonlinear equations is based on the following simple fact:

If an equation $Q(y_1, y_2, \ldots) = 0$ has a solution $y_1 = x_1, y_2 = x_2, \ldots$ in the ring K which satisfies the condition

$$Px_1 = Px_1P, \qquad Px_2 = Px_2P, \ldots, \tag{1.3.1}$$

then the elements $Px_1P, Px_2P, \ldots$ also solve the equation in the subring PKP.

PROOF: Since the element $P = P^2$ is constant by the above condition, then the following identities

$$P\partial_{k_i}^{m_i}(y) \equiv \partial_{k_i}^{m_i}(Py), \qquad \partial_{k_i}^{m_i}(y)P \equiv \partial_{k_i}^{m_i}(yP)$$

are valid, which imply that the elements x_{j_i} meeting the condition (1.3.1) satisfy the equalities

$$P\partial_{k_i}^{m_i}(x_{j_i}) = \partial_{k_i}^{m_i}(Px_{j_i}) = \partial_{k_i}^{m_i}(Px_{j_i}P) = \partial_{k_i}^{m_i}(Px_{j_i}P)P.$$

Hence, if elements $x_1, x_2, \dots \in K$ satisfy the equation $Q(x_1, x_2, \dots) = 0$ and condition (1.3.1), then

$$\begin{aligned}0 = PQ(x_1, x_2, \dots) &= \sum_{j_l} P\partial_{k_1}^{m_1}(x_{j_1})\partial_{k_2}^{m_2}(x_{j_2})\dots A_{j_1,j_2,\dots} \\ &= \sum_{j_l} \partial_{k_1}^{m_1}(Px_{j_1}P)P\partial_{k_2}^{m_2}(x_{j_2})\dots A_{j_1,j_2,\dots}, \\ &= \sum_{j_l} \partial_{k_1}^{m_1}(Px_{j_1}P)\partial_{k_2}^{m_2}(Px_{j_2}P)\dots A_{j_1,j_2,\dots}, \\ &= Q(Px_1P, Px_2P, \dots).\end{aligned}$$

Whence follows validity of the considered statement.

REMARK: Multiplying both sides of the later equality by P from the right, we see that the elements $Px_iP \in PKP$ satisfy also the equation

$$\sum_{j_l} \partial_{k_l}^{m_1}(Px_{j_1}P)\partial_{k_2}^{m_2}(Px_{j_2}P)\dots PA_{j_1,j_2,\dots}P = 0$$

which can be naturally called the projection of the initial equation onto the subring PKP.

Consider now the nonlinear equations of the preceding section which are satisfied by the logarithmic derivatives $\gamma = \Gamma^{-1}\partial\Gamma$. It is important that all these equations admit an additive or multiplicative transformation group. Really, the form of Eqs. (1.2.3), (1.2.5), (1.2.8), (1.2.21) implies that they are satisfied by γ and by $\gamma + c$ as well, where c is an arbitrary constant element. (Here we must regard the operation $[\gamma, B]$ as an inner derivation and, therefore, assume $[c, B] = 0$ if the equation considered contains this operation). So, these equations admit an additive transformation group consisting of all constant elements; if the logarithmic derivative γ satisfies the relation

$$\gamma = \gamma P + N(e - P), \tag{1.3.2}$$

where N and $P = P^2$ are constant elements, then γP satisfies these equations as well. Since $P(\gamma P) = P(\gamma P)P$, the solution γP satisfies the condition (1.3.1) which enables one to apply the projection operation and obtain the solution $P\gamma P$ of the same equation belonging to the subring PKP. Relation (1.3.2) enabling applicability of the projection operation is obviously equivalent to such an equation for Γ:

$$\partial\Gamma(e - P) = \Gamma N(e - P) \tag{1.3.3}$$

which must be adjoined to those linear equations that are simultaneously solved by Γ.

Similarly, the form of Eqs. (1.2.10), (1.2.19), (1.2.20), (1.2.21) implies that they admit a multiplication transformation group consisting of all constant and invertible elements c; it means that they are satisfied by γ and by elements $c\gamma$. Hence, if Γ satisfies the additional equation (1.3.3) while $N \in K^{-1}$, then these equations are satisfied both by γ and simultaneously by the elements

$$N^{-1}\gamma = N^{-1}\gamma P + (e - P)$$

to which the projection operation may be applied, since $PN^{-1}\gamma = PN^{-1}\gamma P$. Thus, elements $PN^{-1}\gamma P$ solve these equations in the subring PKP.

To avoid any ambiguity, we emphasize that, speaking of nonlinear equations, we mean equations in γ, regarding the auxiliary elements u, v, etc. as having been introduced only for brevity. So, strictly speaking, Eq. (1.2.21) is ambiguous: either $u = e + \partial_\alpha(\gamma), \partial u = u\alpha(u) - \alpha^{-1}(u)u$ or $u = \alpha^{-1}(\gamma^{-1})\gamma, \partial u = u\alpha(u) - \alpha^{-1}(u)u$. In the first case it admits an additive transformation group, while in the second case—a multiplicative one. Eq. (1.2.21′) is valid only for the logarithmic derivative $\gamma = \Gamma^{-1}\partial\Gamma$ of the element Γ, the latter satisfying respective linear equations.

Consider, finally, systems of nonlinear equations. Eqs. (1.2.15), (1.2.16) admit, obviously, an additive transformation group $\gamma_1 \to \gamma_1 + c_1, \gamma_2 \to \gamma_2 + c_2$ that is formed by all possible elements c_1, c_2 meeting the conditions $[c_1, D_1] = [c_2, D_2] = 0$. Here the additional condition enabling applicability of the projection operation is expressed by the equations

$$\partial_i\Gamma(e - P) = \Gamma N_i(e - P), \qquad [N_i, D_i] = 0, \tag{1.3.4}$$

and if the condition is met, then elements $P\gamma_i P$ solve Eqs. (1.2.15), (1.2.16) in the subring PKP.

Eq. (1.2.14) admits a multiplication transformation group $\gamma_1 \to \gamma_1 c_1, \gamma_2 \to \gamma_2 c_2$ consisting of invertible elements c_1, c_2 that satisfy the conditions $[c_1, D_2] = [C_2, D_2] = 0$. The additional condition is such:

$$\partial_i\Gamma N_i(e - P) = \Gamma(e - P), \qquad N_i \in K^{-1}, \qquad [N_i, D_i] = 0, \tag{1.3.5}$$

and if it is met, then elements $P\gamma_i N_i P$ solve this equation in the subring PKP.

Concluding the presentation of the general scheme of the method to be suggested, we point out that the ring K_0 is always given from the start, and it is in this ring that one has to look for a solution of the nonlinear equation considered. In the most interesting cases, it is a ring of scalar or matrix functions. The choice of the ring K and the element $P = P^2 \in K$, such that the subring PKP would be coincident (or isomorphic) to K_0, remains at our disposal. Depending on the choice of the ring, we obtain such or other classes of solutions of nonlinear equations considered. Later we shall show that they are wider than those studied by methods of inverse problem and algebraic geometry.

We shall conclude the presentation with a table that sums up the results obtained above. The table is divided into three parts, depending on which of the additional conditions: (1.3.3), (1.3.4), (1.3.5), enabling the projection operation, is met by the element $\Gamma \in K^{-1}$. The additional equation and the formula that solves the appropriate nonlinear equations by means of the projection operation in the subring PKP is written on top of each part. In the left column of the table we write the linear equations satisfied by Γ, while in the right column we write the respective nonlinear equations satisfied by the logarithmic derivative $\gamma = \Gamma^{-1}\partial\Gamma$. We denote constant elements of the ring K by A, A_i, B, C, N, P; throughout the table we assume that $B^2 = e, P^2 = P, [B, P] = 0$. Some line numbers are primed; it means that a particular case is considered of the general equation having the same number without the prime. Then the left column contains only the condition for the particular case.

Table II is given in the Appendix.

CHAPTER 2

REALIZATION OF GENERAL SCHEME IN MATRIX RINGS AND N-SOLITON SOLUTIONS

§1 Wronsky Matrices

Let K_0 be that ring in which one has to look for the solution of the considered nonlinear equation. The simplest extension of this ring is the ring $K = \mathrm{Mat}_N(K_0)$ with elements $a_{ik} \in K_0$ and conventional operations of matrix addition and multiplication. The initial ring K_0 is identified in the natural way with the subring PKP, where

$$P = \begin{pmatrix} e & 0 & \dots & 0 \\ 0 & 0 & \dots & 0 \\ \vdots & \vdots & \vdots & \vdots \\ 0 & 0 & \dots & 0 \end{pmatrix} \tag{2.1.1}$$

and e is the unity of the ring K_0. Extensions of automorphisms α and g.d.'s ∂ of the ring K_0 onto the ring $\mathrm{Mat}_N(K_0)$ are defined by the equalities

$$\alpha(A) = (\alpha(a_{ik})), \qquad \partial A = (\partial a_{ik}), \tag{2.1.2}$$

and they obviously are also automorphisms and g.d.'s in the ring $\mathrm{Mat}_N(K_0)$. They possess the properties

$$\alpha(P) = P, \qquad \partial P = 0.$$

Note that all automorphisms and g.d.'s of the ring $\mathrm{Mat}_N(K_0)$ possessing these properties are obtained by the above extention.

DEFINITION 2.1.1: Matrices $W \in \mathrm{Mat}_N(K_0)$ of the form

$$W = \begin{pmatrix} \partial^{N-1} f_1 & \partial^{N-2} f_1 & \dots & f_1 \\ \vdots & & & \\ \partial^{N-1} f_N & \partial^{N-2} f_N & \dots & f_N \end{pmatrix} \tag{2.1.3}$$

are called Wronsky matrices and denoted by $W\{\partial; f_1, \dots, f_N\}$.

The Wronsky matrices play an important role in solving nonlinear equations because they meet the condition (1.3.3), allowing us to apply the projection operation. Indeed, the matrices

$$N = \begin{pmatrix} 0 & e & 0 & \dots & 0 \\ 0 & 0 & e & \dots & 0 \\ \vdots & \vdots & \vdots & \vdots & \vdots \\ e & 0 & 0 & \dots & 0 \end{pmatrix}, \qquad N^{-1} = \begin{pmatrix} 0 & 0 & \dots & 0 & e \\ e & 0 & \dots & 0 & 0 \\ \vdots & & & & \vdots \\ 0 & 0 & \dots & e & 0 \end{pmatrix} \tag{2.1.4}$$

are invertible in the ring $\mathrm{Mat}_N(K_0)(NN^{-1} = N^{-1}N = e_N$, where e_N is the unity of the ring $\mathrm{Mat}_N(K_0))$ and, according to Eqs. (2.1.1), (2.1.2), (2.1.3), and (2.1.4),

$$\partial W(e_N - P) = \begin{pmatrix} 0 & \partial^{N-1} f_1 & \dots & \partial f_1 \\ 0 & \partial^{N-1} f_N & \dots & \partial f_N \end{pmatrix} = WN(e_N - P).$$

Thus, all Wronsky matrices satisfy the identity

$$\partial W(e_N - P) = WN(e_N - P) \tag{2.1.5}$$

in which the matrices P and N are defined by Eqs. (2.1.1) and (2.1.4). This identity implies that the logarithmic derivative $\gamma = W^{-1}\partial W$ of the invertible (in the ring $\mathrm{Mat}_N(K_0)$) Wronsky matrix has the following form

$$\gamma = \begin{pmatrix} \gamma_1 & e & 0 & \dots & 0 \\ \gamma_2 & 0 & e & \dots & 0 \\ \vdots & & & & \\ \gamma_N & 0 & 0 & \dots & 0 \end{pmatrix}, \tag{2.1.6}$$

its elements $\gamma_j \in K_0$ being found by the following system of equations

$$\partial^N f_i = \sum_{j=1}^{N} \partial^{N-j}(f_i)\gamma_j \qquad (1 \leqslant i \leqslant N). \tag{2.1.7}$$

We remind the reader that, depending on the form of the nonlinear equation, the projection operation is finding only one of the two elements: either $P\gamma P$ or $PN^{-1}\gamma P$. According to Eqs. (2.1.1), (2.1.4), and (2.1.6), we have for the case under consideration,

$$P\gamma P = \begin{pmatrix} \gamma_1 & 0 & \dots & 0 \\ 0 & 0 & \dots & 0 \\ 0 & 0 & \dots & 0 \end{pmatrix}, \qquad PN^{-1}\gamma P = \begin{pmatrix} \gamma_N & 0 & \dots & 0 \\ 0 & 0 & \dots & 0 \\ 0 & 0 & \dots & 0 \end{pmatrix}, \tag{2.1.8}$$

with these matrices being identified with the elements γ_1, γ_N of the ring K_0. Thus, there is no necessity to solve the system (2.1.7), finding all unknown quantities $\gamma_j (1 \leqslant j \leqslant N)$; it is sufficient to find one of the elements: either γ_1 or γ_N.

If the ring K_0 is commutative, then for a matrix $A \in \mathrm{Mat}_N(K_0)$ the notion of its determinant $\mathrm{Det}\, A$ is introduced in the usual way; such determinants possess well-known properties allowing for one to find solutions of systems of linear equations by Cramer's rule, when $\mathrm{Det}\, A$ is invertible in the ring K_0. Besides, in the commutative case for the usual derivation $\partial \in \mathrm{Der}(I)$, the rules of differentiating determinants by rows and by columns remain valid. Hence we get the following lemma.

LEMMA 2.1.1. *If the ring K_0 is commutative and the determinant of Wronsky matrix $W = W\{\partial; f_1, \dots, f_N\}$ is invertible in this ring, then the matrix W itself is invertible in the ring $\mathrm{Mat}_N(K_0)$, and its logarithmic derivative in Eq. (2.1.6) has the form*

$$\gamma_N = (-1)^{N-1} (\mathrm{Det}\, W)^{-1} \,\mathrm{Det}(\partial W). \tag{2.1.9}$$

Besides, if $\partial \in \mathrm{Der}(I)$, then

$$\gamma_1 = (\mathrm{Det}\, W)^{-1} \partial (\mathrm{Det}\, W). \tag{2.1.10}$$

PROOF: Having applied Cramer's rule for solving Eq. (2.1.7), we find

$$\gamma_N = (\mathrm{Det}\, W)^{-1} \,\mathrm{Det}\, W_N, \qquad \gamma_1 = (\mathrm{Det}\, W)^{-1} \,\mathrm{Det}\, W_1,$$

where

$$\mathrm{Det}\, W_N = \begin{vmatrix} \partial^{N-1} f_1 & \dots & \partial f_1 \partial^N f_1 \\ \partial^{N-1} f_N & \dots & \partial f_N \partial^N f_N \end{vmatrix}, \quad \mathrm{Det}\, W_1 = \begin{vmatrix} \partial^N f_1 \partial^{N-2} f_1 & \dots & f_1 \\ \partial^N f_N \partial^{N-2} f_N & \dots & f_N \end{vmatrix}.$$

Equation (2.1.9) follows from these equalities after an appropriate permutation of colums in the determinant $\mathrm{Det}\, W_N$, while Eq. (2.1.10) is obtained by comparing $\mathrm{Det}\, W_1$ with the derivative $\partial(\mathrm{Det}\, W)$ calculated by differentiating the determinant by columns.

Let $f(n; x) = f(n; x_1, \dots, x_m)$ denote scalar functions defined on the direct product of the set Z of all integers by the m-dimensional Euclidian space R^m. The set of all scalar functions $f(n; x)$, infinitely differentiable with respect to variables $x_i (i = 1, \dots, m)$, obviously forms a commutative ring with respect to usual arithmetic operations; we shall denote this ring by $C^\infty(Z \times R^m)$ or simply by C^∞, when it would not cause any confusion. In the ring $C^\infty(Z \times R^m)$, we define the derivation

$$\partial_i \in \mathrm{Der}(I) \qquad (i = 1, \dots m),$$

an automorphism α, and a g.d. $\partial_\alpha = \alpha - I$ described by the equalities

$$\partial_i f(n; x) = \frac{\partial f(n; x)}{\partial x_i}, \qquad \alpha(f(n; x)) = f(n+1; x)$$
$$\partial_\alpha (f(n; x)) = f(n+1; x) - f(n; x).$$

In this chapter, the role of K_0 is played by the rings $\mathrm{Mat}_r(C^\infty)(r = 1, 2, \ldots)$, and only their simplest extentions

$$K = \mathrm{Mat}_N(K_0) = \mathrm{Mat}_N(\mathrm{Mat}_r(C^\infty))$$

are used, formed by block matrices

$$A = (a_{ik}), \qquad a_{ik} \in \mathrm{Mat}_r(C^\infty) \qquad (i, k = 1, \ldots, N).$$

So one should have convenient criteria for block matrix invertibility in the rings

$$\mathrm{Mat}_N(K_0), \qquad K_0 = \mathrm{Mat}_r(C^\infty).$$

The next section is devoted to such criteria.

§2 Conditions of Invertibility of Some Wronsky Matrices

A square numeric matrix A is invertible if and only if the homogeneous system of linear equations corresponding to the matrix has only a zero solution, which is equivalent to absence of nonzero matrix solutions Y of the equation $AY = 0$ in Y, which is equivalent to the relation $\mathrm{Det}\, A \neq 0$. Whence, using well-known expressions for calculating elements of the inverse (reciprocal) matrix through its determinant, we find that a matrix $A(n; x) \in \mathrm{Mat}_r(C^\infty)$ is invertible in this ring if and only if the equation $A(n; x)Y = 0$ in a numeric matrix Y of the order r has only zero solutions for all values $n \in Z, x \in R^m$. Since the ring of block matrices $\mathrm{Mat}_N(\mathrm{Mat}_r(C^\infty))$ is obviously isomorphic to the ring $\mathrm{Mat}_{Nr}(C^\infty)$, the necessary and sufficient condition for the block matrix $A \in \mathrm{Mat}_N(\mathrm{Mat}_r(C^\infty))$, where

$$A = (a_{ik}(n; x)), \qquad a_{ik}(n; x) \in \mathrm{Mat}_r(C^\infty), \qquad i, k = 1, \ldots N$$

to be invertible in this ring is the absence of nonzero solutions of homogeneous systems of equations

$$\sum_{k=1}^{N} a_{ik}(n; x) P_k = 0 \qquad (i = 1, \ldots, N)$$

in numeric matrices P_k of the order r for all values of $n \in Z$, $x \in R^m$. It is in this form that the criterion of invertibility of block matrices in rings $\mathrm{Mat}_N(\mathrm{Mat}_r(C^\infty))$ is used below for proving invertibility-sufficient conditions that are convenient for application.

Let $\{\alpha_i(\Delta), \beta_i(\Delta)\}(1 \leqslant i \leqslant n)$ denote the partition of a half-interval $\Delta = (a, b]$ into parts,

$$\alpha_i(\Delta) = (a_i, a_{i-1}], \qquad \beta_i(\Delta) = (b_{i-1}, b_i]$$

where

$$\begin{aligned} -\infty \leqslant a = a_n < a_{n-1} < \cdots < a_1 < a_0 \\ = b_0 < b_1 < \cdots < b_{n-1} < b_n = b \neq \infty, \end{aligned}$$

and let $d\sigma(\xi)$ denote a nonnegative measure satisfying the inequalities

$$\int_{\alpha_i(\Delta)} d\sigma(\xi) + \int_{\beta_i(\Delta)} d\sigma(\xi) > 0 \qquad (i = 1, \ldots, n)$$

$$\tag{2.2.1} \int_{-\infty}^{\infty} |\xi|^p e(\xi) d\sigma(\xi) < \infty \qquad (p = 1, 2, \ldots),$$

where $e(\xi)$ is a positive $(e(\xi) > 0)$ continuous function.

LEMMA 2.2.1. *If a nonnegative measure $d\sigma(\xi)$ satisfies inequalities (2.2.1), and if for a real polynomial $P(\xi)$ the equality*

$$\tag{2.2.2} \int_{\alpha_i(\Delta)} P(\xi)e(\xi)d\sigma(\xi) + (-1)^{i-1} \int_{\beta_i(\Delta)} P(\xi)e(\xi)d\sigma(\xi) = 0$$

is valid for all $i = 1, \ldots, n$, then this polynomial has at least i roots on every half-interval $(a_i, b_i]$.

PROOF: Substituting $i = 1$ in (2.2.1) and (2.2.2), we find

$$\int_{(a_1,b_1]} d\sigma(\xi) > 0, \qquad \int_{(a_1,b_1]} P(\xi)e(\xi)d\sigma(\xi) = 0,$$

whence, due to the nonnegativity of the measure $d\sigma(\xi)$ and strict positivity of the function $e(\xi)$, it follows that at least one root of the polynomial lies in the half-interval $(a_1, b_1]$. Thus, the statement to be proved is true for $i = 1$. Suppose it is not true for all i's, and let i_0 be the first number for which the statement is false. Then: 1) $i_0 > 1$, 2) the half-interval $(a_{i_0-1}, b_{i_0-1}]$ contains exactly $i_0 - 1$ roots of the polynomial $P(\xi)$; and 3) the set $\alpha_{i_0}(\Delta) \cup \beta_{i_0}(\Delta)$ contains no roots of the polynomial. Therefore, on all half-intervals $\alpha_{i_0}(\Delta), \beta_{i_0}(\Delta)$ the polynomial preserves its sign, with the sign being the same, if the root number $i_0 - 1$ on the half-interval $(a_{i_0-1}, b_{i_0-1}]$ is even, and with the sign being opposite, if the root number is odd. Thus, the sign of $P(\xi)$ on the half-interval $\alpha_{i_0}(\Delta)$ coincides with the sign of $(-1)^{i_0-1}P(\xi)$ on the half-interval $\beta_{i_0}(\Delta)$. On the other hand, substituting $i = i_0$ in (2.2.1) and (2.2.2), we get

$$\int_{\alpha_{i_0}(\Delta)} d\sigma(\xi) + \int_{\beta_{i_0}(\Delta)} d\sigma(\xi) > 0,$$

$$\int_{a_{i_0}(\Delta)} P(\xi)e(\xi)d\sigma(\xi) + \int_{\beta_{i_0}(\Delta)} (-1)^{i_0-1}P(\xi)e(\xi)d\sigma(\xi) = 0.$$

Since the measure $d\sigma(\xi)$ is nonnegative and the sign of the function $P(\xi)e(\xi)$ on the half-interval $\alpha_{i_0}(\Delta)$ coincides with the sign of $(-1)^{i_0-1}P(\xi)e(\xi)$ on $\beta_{i_0}(\Delta)$, these two relationships cannot be valid simultaneously, and therefore the assumption leads to contradiction.

Now let $M = \bigcup_{k=1}^{l} \Delta_k$ be a union of mutually disjoint half-intervals Δ_k, let $\{\alpha_i(\Delta_k), \beta_i(\Delta_k)\}(1 \leqslant i \leqslant n_k)$ be a partition of a half-interval Δ_k, and let $\varphi(\xi)d\sigma(\xi)$ be a continuous function defined on the set M and a nonnegative measure, respectively. From Lemma 2.2.1 we deduce the following.

COROLLARY 2.2.1. *If* $\varphi(\xi) > 0$,

$$\int\limits_{\alpha_i(\Delta_k)} d\sigma(\xi) + \int\limits_{\beta_i(\Delta_k)} d\sigma(\xi) > 0 \qquad (1 \leqslant i \leqslant n_k, 1 \leqslant k \leqslant l),$$

and for all $x \in R^1$,

$$\int\limits_M e^{\xi x}\varphi(\xi)d\sigma(\xi) < \infty, \tag{2.2.3}$$

then the functions

$$f_i^{\Delta_k}(x) = \int\limits_{\alpha_i(\Delta_k)} e^{\xi x}\varphi(\xi)d\sigma(\xi) + (-1)^{i-1}\int\limits_{\beta_i(\Delta_k)} e^{\xi x}\varphi(\xi)d\sigma(\xi) \tag{2.2.4}$$

$(1 \leqslant i \leqslant n_k, 1 \leqslant k \leqslant l)$ *are infinitely differentiable, the matrix*

$$W = W\left\{\partial; f_1^{\Delta_1}, \ldots, f_{n_1}^{\Delta_1}, \ldots, f_1^{\Delta_l}, \ldots, f_{n_l}^{\Delta_l}\right\} (\partial = \frac{\partial}{\partial x})$$

belongs to the ring $\mathrm{Mat}_N(C^\infty)(N = \sum n_k)$ *and is invertible in it.*

PROOF: The fact that the functions $f_i^{\Delta_k}(x)$ are infinitely differentiable and the matrices W belong to the ring $\mathrm{Mat}\, N(C^\infty)$ immediately follows from the inequalities (2.2.3) and positivity of the function $\varphi(\xi)$. To demonstrate the invertibility of the matrix W in this ring, one must prove that for every $x \in R^1$ the system of homogeneous linear equations

$$\sum_{j=1}^{N} \partial^{N-j}(f_i^{\Delta_k}(x))p_j = 0,$$

$$(1 \leqslant i \leqslant n_k, 1 \leqslant k \leqslant l, 1 \leqslant j \leqslant N = n_1 + n_2 + \cdots + n_e)$$

in unknown quantities p_j has solely the zero solution, and, since the matrix W is real, only real solutions may be considered. According to Eq. (2.2.4), this system of equations is equivalent to

$$\int\limits_{\alpha_i(\Delta_k)} P_{N-1}(\xi)e^{\xi x}\varphi(\xi)d\sigma(\xi) + (-1)^{i-1}\int\limits_{\beta_i(\Delta_k)} P_{N-1}(\xi)e^{\xi x}\varphi(\xi)d\sigma(\xi) = 0,$$

where $P_{N-1}(\xi) = \sum_{j=1}^{N} \xi^{N-j}p_j$ is a real polynomial whose power does not exceed $N-1$. From Lemma 2.2.1 and the positivity of the function $e^{\xi x}\varphi(\xi)$ one can deduce that these equalities imply existence of not less than n_k roots of the polynomial $P_{N-1}(\xi)$ in every half-interval Δ_k. All in all, the polynomial has not less than $n_1 + n_2 + \cdots + n_l = N$ roots, which is possible only if $p_1 = p_2 = \cdots = p_N = 0$.

REMARK: Besides ξ the function φ can depend on the variables $n \in Z$, $y \in R^{m-1}$: $\varphi = \varphi(\xi; n, y)$. If it is infinitely differentiable with respect to y and if partial derivatives of any order of the function $e^{\xi x}\varphi(\xi; n, y)$ with respect to the variables x, y are summable on M in the measure $d\sigma(\xi)$, then the matrix $W = W(n; x, y)$ belongs to the ring $\mathrm{Mat}_N(C^\infty(Z \times R^m))$ and is invertible in it.

We shall call a set of complex numbers Λ *symmetrical* if $\lambda \in \Lambda$ implies $\bar\lambda \in \Lambda$, where $\bar\lambda$ is the complex conjugate to λ.

LEMMA 2.2.2.. *Let $\{\lambda_1, \lambda_2, \ldots, \lambda_N\}$ be a symmetrical set of complex numbers that lie in the open right half-plane ($\mathrm{Re}\,\lambda_k > 0$) and let $u(\lambda)$ and $v(\lambda)$ be two functions defined on the set and satisfying the conditions*

$$u(\bar\lambda_k) = \overline{u(\lambda_k)}, \qquad v(\bar\lambda_k) = \overline{v(\lambda_k)} \\ |u(\lambda_k)| + |v(\lambda_k)| > 0. \tag{2.2.5}$$

Then the matrix $C = (c_{ij})(i, j = 1, 2, \ldots, N)$ with the elements

$$c_{ij} = \lambda_i^{N-j} u(\lambda_i) + i(-\lambda_i)^{N-j} v(\lambda_i) \tag{2.2.6}$$

has a nonzero determinant.

PROOF: It is sufficient to show that the system of linear homogeneous equations

$$\sum_{j=1}^{N} c_{ij} p_j = 0 \qquad (i = 1, 2, \ldots, N),$$

with respect to $p_1, p_2, \ldots, p_N$, has no nonzero solutions. Equation (2.2.6) implies that these equations are equivalent to the following ones,

$$u(\lambda_i)P_{N-1}(\lambda_i) + iv(\lambda_i)P_{N-1}(-\lambda_i) = 0 \qquad (1 \leqslant i \leqslant N), \tag{2.2.7}$$

where $P_{N-1}(\lambda) = \sum_{j=1}^{N} \lambda^{N-j} p_j$ is the polynomial of the power $\leqslant N - 1$. Since the set $\{\lambda_i\}$ is symmetrical, the equalities (2.2.7) are satisfied together with the following ones,

$$u(\bar\lambda_i)P_{N-1}(\bar\lambda_i) + iv(\bar\lambda_i)P_{N-1}(-\bar\lambda_i) = 0,$$

while from Eq. (2.1.5) they must be equivalent to

$$u(\lambda_i)\bar P_{N-1}(\lambda_i) - iv(\lambda_i)\bar P_{N-1}(-\lambda_1) = 0, \tag{2.2.7'}$$

where $\bar P_{N-1}(\lambda) = \sum_{j=1}^{N}, \lambda^{N-j}\bar p_j$. Since, by assumption, one of the quantities $u(\lambda_i), v(\lambda_i)$ differs from zero for every i, Eqs. (2.2.7) and (2.2.7′) can be simultaneously satisfied only if

$$P_{N-1}(\lambda_i)\bar P_{N-1}(-\lambda_i) + \bar P_{N-1}(\lambda_i)P_{N-1}(-\lambda_i) = 0.$$

Hence the polynomial

$$Q(\lambda) = P_{N-1}(\lambda)\bar{P}_{N-1}(-\lambda) + \bar{P}_{N-1}(\lambda)P_{N-1}(-\lambda)$$

of the power $\leqslant 2(N-1)$ has N roots in the open right half-plane: $\lambda_1, \lambda_2, \dots, \lambda_N$; and since the polynomial has even power, its other N roots lie in the left half-plane: $-\lambda_1, -\lambda_2, \dots, -\lambda_N$, which can happen only when $Q(\lambda) \equiv 0$. But on the imaginary axis $\lambda = i\tau$,

$$0 \equiv Q(\lambda) = |P_{N-1}(i\tau)|^2 + |P_{N-1}(-i\tau)|^2,$$

which implies that $P_{N-1}(\lambda) \equiv 0$, whence $p_1 = p_2 = \cdots = p_N = 0$.

COROLLARY 2.2.2. *Let a symmetrical set of complex numbers* $\{\lambda_1, \lambda_2, \dots, \lambda_N\}$ *lying in the open right half-plane* ($\operatorname{Re}\lambda_k > 0$) *contain* p *real and* r *pairs of complex conjugate numbers* ($p+2r = N$). *If the functions* $u(\lambda, y), v(\lambda, y) (y \in R^{m-1})$ *satisfy the conditions*

$$u(\bar{\lambda}, y) = \overline{u(\lambda, y)}, \qquad v(\bar{\lambda}, y) = \overline{v(\lambda, y)}$$
$$|u(\lambda, y)| + |v(\lambda, y)| > 0$$

and are infinitely differentiable with respect to y, *then the Wronsky matrix* $W = W\{\partial; f_1, f_2, \dots, f_N\}$, *where* $\partial = \frac{\partial}{\partial x}$, *and*

$$f_j = f_j(x, y) = e^{\lambda_j x}u(\lambda_j, y) + ie^{-\lambda_j x}(\lambda_j, y)$$

belongs to the ring $\operatorname{Mat}_N(C^\infty)$ *and is invertible in it. Besides*

$$\partial W = \Lambda M\overline{W}, \tag{2.2.8}$$

where $\overline{W}$ *is the matrix whose elements are complex conjugate with the appropriate elements of* W; M *is the matrix transposing the rows corresponding to the conjugate values of* λ_j *and leaving all other rows in their former positions; and* $\Lambda = \operatorname{diag}\{\lambda_1, \lambda_2, \dots, \lambda_N\}$ *is a diagonal matrix with the elements* $\lambda_1, \lambda_2, \dots, \lambda_N$ *on its principal diagonal.*

In addition, in formula (2.1.6) for the logarithmic derivative $\gamma = W^{-1}\partial W$,

$$\gamma_N = \gamma_N(x, y) = (-1)^{N-1+r}\Big(\prod_{j=1}^{N}\lambda_j\Big)\exp(-2i\arg\operatorname{Det}W).$$

PROOF: By definition, the Wronsky matrix $W = W\{\partial; f_1, \dots, f_N\}$ has the elements

$$w_{jk} = \partial^{N-k}(f_j) = (\lambda_j)^{N-k}e^{\lambda_j x}u(\lambda_j, y) + i(-\lambda_j)^{N-k}e^{\lambda_j x}v(\lambda_j, y).$$

They obviously belong to $C^\infty(R^m)$ and satisfy all conditions of Lemma 2.2.2 for any values of $x \in R^1$, $y \in R^{m-1}$. That is why $\det W \neq 0$, for all x, y, and the matrix W belongs to the ring $\mathrm{Mat}_N(C^\infty)$ and is invertible in it. Further, the elements w_{jk} of the matrix ∂W are given by

$$\begin{aligned} w'_{jk} &= \lambda_j\{(\lambda_j)^{N-k}e^{\lambda_j x}u(\lambda_j,y) - i(-\lambda_j)^{N-k}e^{-\lambda_j x}v(\lambda_j,y)\} \\ &= \lambda_j\overline{\{(\bar\lambda_j)^{N-k}e^{\bar\lambda_j x}\overline{u(\lambda_j,y)} + i(-\bar\lambda_j)^{N-k}e^{-\bar\lambda_j x}\overline{v(\lambda_j,y)}\}}, \end{aligned}$$

whence, using the equalities

$$\overline{u(\lambda_j,y)} = u(\bar\lambda_j,y), \qquad \overline{v(\lambda_j,y)} = v(\bar\lambda_j,y),$$

we find

$$w'_{jk} = \lambda_j\bar w_{j',k},$$

where by j' we denote the number for which $\lambda_{j'} = \bar\lambda_j$. Equation (2.2.8) immediately follows from these equalities. Finally, according to Eq. (1.1.9), $\gamma_N = (-1)^{N-1}(\mathrm{Det}\,W)^{-1}\,\mathrm{Det}(\partial W)$, whence, using Eq. (2.2.8) that has been just proved, we find that

$$\begin{aligned} \gamma_N &= (-1)^{N-1}\,\mathrm{Det}\,\Lambda\,\mathrm{Det}\,M(\mathrm{Det}\,W)^{-1}(\mathrm{Det}\,\overline{W}) \\ &= (-1)^{N-1+r}(\prod_{j=1}^{N}\lambda_j\exp(-2i\arg\mathrm{Det}\,W). \end{aligned}$$

Until now we considered a commutative ring $K_0 = C^\infty$. Now we shall introduce a noncommutative case where $K_0 = \mathrm{Mat}_r(C^\infty)$.

LEMMA 2.2.3. *Let $\{\lambda_1,\lambda_2,\ldots,\lambda_N\}$ be an arbitrary set of complex numbers lying in the open right half-plane ($\mathrm{Re}\,\lambda_k > 0$), I be a unit matrix of the rth order, and $Q_1, Q_2 = (I - Q_1) \in \mathrm{Mat}_r$ be constant orthoprojectors ($Q_i = Q_i^* = Q_i^2, i = 1,2$). Then the Wronsky matrix $W = W\{\partial; f_1,\ldots,f_N\}$, where $\partial = \frac{\partial}{\partial x}$,*

$$f_j = e^{\lambda_j x}u_j(t)Q_1 + e^{-\lambda_j x}B_j(t) - e^{-\bar\lambda_j x}B_j(t)^* + e^{\bar\lambda_j x}\overline{u_j(t)}Q_2, \tag{2.2.9}$$

$u_j(t) \in C^\infty$ are scalar functions, $B_j(t) \in \mathrm{Mat}_r(C^\infty)$ are matrix functions satisfying the conditions

$$u_j(t) \neq 0, \qquad Q_1B_j(t)Q_2 = B_j(t),$$

belongs to the ring $\mathrm{Mat}_N(K_0)$ ($K_0 = \mathrm{Mat}_r(C^\infty)$) and is invertible in it.

PROOF: It will suffice to prove that for any fixed $x \in R^1$, $t \in R^1$, the system of linear homogeneous equations

$$\sum_{k=1}^{N}\partial^{N-k}(f_j)p_k = 0 \qquad (1 \leqslant j \leqslant N),$$

with respect to matrices of the rth order $p_1, p_2, \ldots, p_N$ has no nonzero solutions. From Eq. (2.2.9) it follows that the equations are equivalent to

$$e^{\lambda_j x} u_j(t) Q_1 P_{N-1}(\lambda_j) + e^{-\lambda_j x} B_j(t) P_{N-1}(-\lambda_j) \\ - e^{-\bar{\lambda}_j x} B_j(t)^* P_{N-1}(-\bar{\lambda}_j) \\ + e^{\bar{\lambda}_j x} \overline{u_j(t)} Q_2 P_{N-1}(\bar{\lambda}_j) = 0,$$

where $P_{N-1}(\lambda) = \sum_{k=1}^{N} \lambda^{N-k} p_k$ is a polynomial of the power $\leqslant N-1$ with matrix coefficients $p_k \in \mathrm{Mat}_r$. Multiplying these equations by the matrices Q_1, Q_2 from the left and noting that by assuming

$$Q_1, Q_2 = Q_2 Q_1 = 0, \qquad Q_1 B_j = B_j, \qquad Q_2 B_j^* = B_j^*,$$

we see that they fall into the following pairs of equations,

$$e^{\lambda_j x} u_j(t) Q_1 P_{N-1}(\lambda_j) + e^{-\lambda_j x} B_j(t) P_{N-1}(-\lambda_j) = 0,$$
$$-e^{-\bar{\lambda}_j x} B_j(t)^* P_{N-1}(-\bar{\lambda}_j) + e^{\bar{\lambda}_j x} \overline{u_j(t)} Q_2 P_{N-1}(\bar{\lambda}_j) = 0.$$

Using conjugate matrices in the latter equation, we obtain the equations

$$e^{\lambda_j x} u_j(t) Q_1 P_{N-1}(\lambda_j) + e^{-\lambda_j x} B_j(t) P_{N-1}(-\lambda_j) = 0,$$
$$-e^{-\lambda_j x} P_{N-1}^*(-\lambda_j) B_j(t) + e^{\lambda_j x} u_j(t) P_{N-1}^*(\lambda_j) Q_2 = 0,$$

which are equivalent to the initial ones. Here

$$P_{N-1}^*(\lambda) = \sum_{k=1}^{N} \lambda^{N-k} p_k^* = (P_{N-1}(\bar{\lambda}))^*.$$

Multiply the first equation by $P_{N-1}^*(-\lambda_j)$ from the left and the second equation by $P_{N-1}(-\lambda_j)$ from the right. Summing the products, we arrive at the following equalities,

$$e^{\lambda_j x} u_j(t) \{ P_{N-1}^*(-\lambda_j) Q_1 P_{N-1}(\lambda_j) + P_{N-1}^*(\lambda_j) Q_2 P_{N-1}(-\lambda_j) \} = 0.$$

Since $e^{\lambda_j x} u_j(t) \neq 0$ by the assumption, these equalities imply that the polynomial

$$R(\lambda) = P_{N-1}^*(-\lambda) Q_1 P_{N-1}(\lambda) + P_{N-1}^*(\lambda) Q_2 P_{N-1}(-\lambda)$$

of the power $\leqslant 2N-2$ has N roots $\lambda_1, \lambda_2, \ldots, \lambda_N$ in the open right half-plane. Since $(R(\lambda))^* = R(-\bar{\lambda})$, it has N more roots $-\bar{\lambda}_1, -\bar{\lambda}_2, \ldots, -\bar{\lambda}_N$ in the left half-plane which is possible only if $R(\lambda) \equiv 0$. Finally, setting $\lambda = i\tau (-\infty < \tau < \infty)$, we find

$$P_{N-1}(i\tau)^* Q_1 P_{N-1}(i\tau) + P_{N-1}(-i\tau)^* Q_2 P_{N-1}(-i\tau) \equiv 0,$$
$$P_{N-1}(-i\tau)^* Q_1 P_{N-1}(-i\tau) + P_{N-1}(i\tau)^* Q_2 P_{N-1}(i\tau) \equiv 0,$$

and, therefore,

$$P_{N-1}(i\tau)^* P_{N-1}(i\tau) + P_{N-1}(-i\tau)^* P_{N-1}(-i\tau) \equiv 0,$$

which is possible only if $p_1 = p_2 = \cdots = p_N = 0$.

COROLLARY 2.2.3. *If the conditions of Lemma 2.2.3 hold, then the logarithmic derivative $\gamma = W^{-1}\partial W$ obeys Eq. (2.1.6) with*

$$[\gamma_1, B]^* = [\gamma_1, B], \qquad (\tilde{\gamma}_N)^* = (\tilde{\gamma}_N)^{-1} \tag{2.2.10}$$

where

$$B = Q_1 - Q_2, \qquad \tilde{\gamma}_N = \left(\prod_{j=1}^{N} |\lambda_j|\right)^{-1} \gamma_N.$$

Only Eq. (2.2.10) needs to be proved since Eq. (2.1.6) is valid for all invertible Wronsky matrices. Repeating the calculations used in the proof of Lemma 2.2.3, we see that from Eqs. (2.1.7) which are satisfied by the matrices $\gamma_i \in \mathrm{Mat}_r(C^\infty)$ it follows that the polynomial of the power $2N$

$$T(\lambda) = P_N^*(-\lambda)Q_1P_N(\lambda) + P_N^*(\lambda)Q_2P_N(-\lambda),$$

where

$$P_N(\lambda) = -(\lambda)^N I + \sum_{k=1}^{N} \lambda^{N-k}\gamma_k, \qquad P_N^*(\lambda) = (P_N(\bar{\lambda}))^*$$

equals zero when $\lambda = \lambda_1, \lambda_2, \ldots, \lambda_N, -\bar{\lambda}_1, -\bar{\lambda}_2, \ldots, -\bar{\lambda}_N$. Therefore, $T(\lambda) = C\prod_{1\leqslant j\leqslant N}(\lambda-\lambda_j)(\lambda+\bar{\lambda}_j)$, where the matrix $C \in \mathrm{Mat}_r(C^\infty)$ and is independent of λ. On the other hand, the definition of this polynomial and the equalities $Q_1 + Q_2 = I, B = Q_1 - Q_2$ immediately imply that

$$T(\lambda) = (-1)^N\{\lambda^{2N}I + \lambda^{2N-1}(\gamma_1^*B - B\gamma_1) + \cdots + \gamma_N^*\gamma_N\}.$$

That is why $C = (-1)^N I$ and

$$\left(\sum_{j=1}^{N}(\bar{\lambda}_j - \lambda_j)\right) I = \gamma_1^*B - B\gamma_1, \qquad \left(\prod_{j=1}^{N} \lambda_j\bar{\lambda}_j\right) I = \gamma_N^*\gamma_N.$$

Since $B^2 = I$, then after multiplying the first of these equalities by B on the left and on the right, we find that $\gamma_1^*B - B\gamma_1 = B\gamma_1^* - \gamma_1B$, i.e., $[\gamma_1, B] = -[\gamma_1^*, B] = [\gamma_1, B]^*$. Having divided both sides of the second equality by $\prod_{j=1}^N |\lambda_j|^2$, we obtain $\tilde{\gamma}_N^*\tilde{\gamma}_N = I$ and, hence, $(\tilde{\gamma}_N)^* = (\tilde{\gamma}_N)^{-1}$, where $\tilde{\gamma}_N = (\prod_{j=1}^N |\lambda_j|)^{-1}\gamma_N$.

Consider a block matrix

$$C = (c_{ik})(c_{ik} \in \mathrm{Mat}_r, i, k = 1, 2, \ldots, N)$$

with elements

$$c_{ik} = Q_1(i)\lambda_i^{N-k}\mathcal{E}(i) + Q_2(i)(-\bar{\lambda}_i)^{N-k}\mathcal{E}(i)^{*-1}, \tag{2.2.11}$$

where $\mathcal{E}(i) \in \mathrm{Mat}_r$ are arbitrary invertible matrices and

$$Q_1(i), Q_2(i) = I - Q_1(i) \in \mathrm{Mat}_r$$

are arbitrary orthoprojectors

$$(Q_1(i)^2 = Q_1(i) = Q_1(i)^*).$$

LEMMA 2.2.4. *If numbers $\lambda_i (1 \leqslant i \leqslant N)$ are different and lie in the open right half-plane, then the matrix C is invertible.*

PROOF: Formulas (2.2.11) imply that the system of homogeneous linear equations

$$\sum_{k=1}^{N} c_{ik} p_{N-k} = 0 \qquad (1 \leqslant i \leqslant N) \tag{2.2.12}$$

in matrices $p_0, p_1, \ldots, p_{N-1}$ of the order r is equivalent to the equations

$$Q_1(i)\mathcal{E}(i)P_{N-1}(\lambda_i) + Q_2(i)\mathcal{E}(i)^{*-1}P_{N-1}(-\bar{\lambda}_i) = 0 \qquad (1 \leqslant i \leqslant N),$$

where $P_{N-1}(z) = \sum_{j=1}^{N} z^{N-j} p_{N-j}$ is a polynomial of the power not exceeding $N-1$ with matrix coefficients $p_{N-1} \in \mathrm{Mat}_r$. Multiplying these equations by the orthoprojectors $Q_1(i), Q_2(i)$ from the left, we see that

$$Q_1(i)\mathcal{E}(i)P_{N-1}(\lambda_i) = 0, \qquad Q_2(i)\mathcal{E}(i)^{*-1}P_{N-1}(-\bar{\lambda}_i) = 0,$$

and since $Q_1(i) + Q_2(i) = I$, then,

$$\mathcal{E}(i)P_{N-1}(\lambda_i) = Q_2(i)\mathcal{E}(i)P_{N-1}(\lambda_i),$$
$$\mathcal{E}(i)^{*-1}P_{N-1}(-\bar{\lambda}_i) = Q_1(i)\mathcal{E}(i)^{*-1}P_{N-1}(-\bar{\lambda}_i).$$

Let us perform the conjugation operation in the first of these equalities. As a result, we obtain the equalities

$$P_{N-1}(\lambda_i)^*\mathcal{E}(i)^* = P_{N-1}(\lambda_i)^*\mathcal{E}(i)^*Q_2(i)^*,$$
$$\mathcal{E}(i)^{*-1}P_{N-1}(-\bar{\lambda}_i) = Q_1(i)\mathcal{E}(i)^{*-1}P_{N-1}(-\bar{\lambda}_i).$$

Multiplying them by one another, we get

$$P_{N-1}(\lambda_i)^*P_{N-1}(-\bar{\lambda}_i) = P_{N-1}(\lambda_i)^*\mathcal{E}(i)^*Q_2(i)^*Q_1(i)\mathcal{E}(i)^{*-1}P_{N-1}(-\bar{\lambda}_i) = 0,$$

since by the condition $Q_2(i)^* = Q_2(i), Q_{2},(i)Q_1(i) = 0$.

Setting

$$P_{N-1}^*(z) = \sum_{j=1}^{N} z^{N-j} p_{N-j}^*, \qquad Q_{2N-2}(z) = P_{N-1}^*(z)P_{N-1}(-z),$$

we conclude that the polynomial $Q_{2N-2}(z)$ of the power $2N-2$ (with matrix coefficients) vanishes at $z = \bar{\lambda}_1, \bar{\lambda}_2, \ldots, \bar{\lambda}_N$, and since

$$\begin{aligned} Q_{2N-2}(z)^* &= P_{N-1}(-z)^*(P_{N-1}^*(z))^* \\ &= P_{N-1}^*(-\bar{z})P_{N-1}(\bar{z}) = Q_{2N-2}(-\bar{z}), \end{aligned}$$

the polynomial equals zero also at $z = -\lambda_1, -\lambda_2, \ldots, -\lambda_N$. By the condition, all numbers $\{\bar{\lambda}_1, \bar{\lambda}_2, \ldots, \bar{\lambda}_N, -\bar{\lambda}_1, -\bar{\lambda}_2, \ldots, -\bar{\lambda}_N\}$ are different, and so it follows that the polynomial $Q_{2N-2}(z)$ is equal to zero. Therefore the identity

$$Q_{2N-2}(iy) = P_{N-1}^*(iy)P_{N-1}(-iy) = P_{N-1}(-iy)^*P_{N-1}(-iy) \equiv 0$$

is valid on the imaginary axis $z = iy(-\infty < y < \infty)$, which is possible only if $p_0 = p_1 = \cdots = p_{N-1} = 0$.

Thus, the homogeneous system (2.2.12) has only trivial solutions, and the matrix C is invertible.

COROLLARY 2.2.4. *If the conditions of Lemma 2.2.4 are met and the matrices* $u_1, u_2, \ldots, u_N \in \mathrm{Mat}_r$ *are found from the simultaneous equations*

$$\sum_{k=1}^{N} c_{ik} u_k = c_{i0} \qquad (1 \leqslant i \leqslant N),$$

then

$$(u_1 + i\alpha I)^* = (u_1 + i\alpha I), \qquad u_N^* u_N = \rho^2 I,$$

where

$$\alpha = \sum_{k=1}^{N} \mathrm{Im}\,\lambda_k, \qquad \rho = \prod_{k=1}^{N} |\lambda_k|.$$

These statements are proved like Corollary 2.2.3.

§3 N-Soliton Solutions of Nonlinear Equations

1 *KdV and Nonlinear String Equations*

Solutions of these equations are sought in the ring $K_0 = C^\infty$ of scalar infinitely differentiable functions $f(x,t)$ with standard differentiation

$$\partial_0 f = \frac{\partial f}{\partial t}, \qquad \partial f = \frac{\partial f}{\partial x}.$$

As an extention of the ring, we take the ring $K = \mathrm{Mat}_N(C^\infty)$ of infinitely differentiable square matrix functions of the order N.

Comparing formula 1 of Table II with the results described in §1 of the present chapter, we find that if the invertible Wronsky matrix

$$W = W\{\partial; f_1, \ldots, f_N\} \in \mathrm{Mat}_N(C^\infty)$$

satisfies the equations

$$(\partial_0 + \partial^3)W = 0, \qquad \partial^2 W = AW, \tag{2.3.1}$$

where $A \in \mathrm{Mat}_N$ is a constant matrix, then the function

$$v = -\frac{\partial^2}{\partial x^2} \ln \mathrm{Det}\, W \tag{2.3.2}$$

satisfies the equation

$$4v_t + v_{xxx} - 12 v v_x = 0. \tag{2.3.3}$$

If A is a diagonal matrix ($A = \mathrm{diag}(\mu_1^2, \mu_2^2, \ldots, \mu_N^2)$), then Eqs. (2.3.1) will split into similar equations for f_j

$$\frac{\partial f_j}{\partial t} + \frac{\partial^3 f_j}{\partial x^3} = 0, \qquad \frac{\partial^2 f_j}{\partial x^2} = \mu_j^2 f_j.$$

The general solution of these equations is given by the functions

$$f_j = e^{\mu_j x - \mu_j^3 t} C_j(1) + e^{-(\mu_j x - \mu_j^3 t)} C_j(2),$$

where $C_j(1), C_j(2)$ are arbitrary numbers. Setting

$$0 < \mu_1 < \mu_2 < \cdots < \mu_N,$$
$$2C_j(1) = e^{m_j}, \qquad 2C_j(2) = (-1)^{j-1} e^{-m_j}, \qquad (\operatorname{Im} m_j = 0),$$

we obtain the real functions

$$f_j = \frac{1}{2}(e^{\theta_j} + (-1)^{j-1} e^{-\theta_j}), \qquad \theta_j = \mu_j x - \mu_j^3 t + m_j$$

that satisfy the conditions of Corollary 2.2.1 which guarantee the invertibility of the matrix $W = W\{\partial; f_1, \ldots, f_N\}$ in the ring $\mathrm{Mat}_N(C^\infty)$. Therefore, after so choosing the parameters $\mu_j, C_j(1), C_j(2)$, the functions $v = v(x,t)$ defined by (2.3.2) are real infinitely differentiable solutions of equation (2.3.3). They depend on $2N$ real parameters μ_j, m_j and are called N-soliton solutions. Compact formulas for these solutions have obviously the form:

$$W = W\{\partial; \operatorname{ch}\theta_1, \operatorname{sh}\theta_2, \operatorname{ch}\theta_3, \ldots\}$$
$$\theta_j = \mu_j x - \mu_j^3 t + m_j, \qquad 0 < \mu_1 < \mu_2 < \cdots < \mu_N, \qquad \operatorname{Im} m_j = 0$$
$$v(x,t) = -\frac{\partial^2}{\partial x^2} \ln \operatorname{Det} W.$$

Similarly, formula II (1″) implies that if the invertible Wronsky matrix $W = W\{\partial; f_1, \ldots, f_N\}$ satisfies the equations

$$(\lambda\partial + \partial^3)W = DW, \qquad (\varepsilon\partial_0 + \partial^2)W = AW, \tag{2.3.4}$$

($D, A \in \mathrm{Mat}_N$ are constant matrices, λ, ε are numbers), then the function

$$v = -\frac{\partial u}{\partial x} = -\frac{\partial^2}{\partial x^2} \ln \operatorname{Det} W \tag{2.3.5}$$

satisfies the equation

$$4\lambda v_{xx} + 3\varepsilon^2 v_{tt} + (v_{xx} - 6v^2)_{xx} = 0. \tag{2.3.6}$$

(The term $6[\partial_1 u, \partial u]$ vanishes due to commutativity of the ring $k_0 = C^\infty$). If

$$D = \operatorname{diag}(d_1, d_2, \ldots, d_N), \qquad A = \operatorname{diag}(a_1, a_2, \ldots, a_N),$$

then Eqs. (2.3.4) split into similar equations for the functions f_j

$$(\lambda\partial + \partial^3) f_j = d_j f_j, \qquad (\varepsilon\partial_0 + \partial^2) f_j = a_j f_j, \tag{2.3.7}$$

whose general solutions have the form

$$f_j = e^{\varepsilon^{-1}a_j t} \sum_{p=1}^{3} e^{z_j(p)x - \varepsilon^{-1} z_j(p)^2 t} C_j(p), \tag{2.3.8}$$

where $z_j(p)$ are roots of the equations $z^3 + \lambda z = d_j$ and $C_j(p)$ are arbitrai numbers.

Equation (2.3.6) is of interest only for real-valued parameters λ, ε^2 whic can be taken equal to ± 1 without a loss of generality. The problem is to fin real solutions of these equations.

The roots of the equations $z^3 + \lambda z = d_j$ are expressed through one of thei $z_j(1)$, by the formulas

$$z_j(2) = \frac{-z_j(1) - \sqrt{-4\lambda - 3z_j(1)^2}}{2},$$
$$z_j(3) = \frac{-z_j(1) + \sqrt{-4\lambda - 3z_j(1)^2}}{2}$$

which allow us to introduce such a parameterization:

$$z_j(1) = 2i\sqrt{\lambda}\alpha_j,$$
$$z_j(2) = -i\sqrt{\lambda}\,(\alpha_j - \beta_j),$$
$$z_j(3) = -i\sqrt{\lambda}\,(\alpha_j + \beta_j).$$

Here α_j, β_j stand for arbitrary numbers related by the expression

$$3\alpha_j^2 + \beta_j^2 = 1$$

and $d_j = 2i\lambda\sqrt{\lambda}(1 - 4\alpha_j^2)$. Choosing elements a_j of the matrix A to equ $-\lambda(\alpha_j^2 + \beta_j^2)$, we obtain, according to Eq. (2.3.6),

$$\begin{aligned} f_j = {} & e^{2_j\sqrt{\lambda}\,\alpha_j x + \varepsilon^{-1}\lambda(3\alpha_j^2 - \beta_j^2)t} C_j(1) \\ & + e^{-i\sqrt{\lambda}\,(\alpha_j - \beta_j)x - 2\varepsilon^{-1}\lambda\alpha_j\beta_j t} C_j(2) \\ & + e^{-i\sqrt{\lambda}\,(\alpha_j + \beta_j)x + 2\varepsilon^{-1}\lambda\alpha_j\beta_j t} C_j(3). \end{aligned} \tag{2.3.9}$$

Let $\lambda = -1, \varepsilon = 1$. We shall fit the parameters in Eqs. (2.3.9) so that the matri $W\{\partial; f_1, \dots, f_N\}$ becomes real and invertible in the ring $\mathrm{Mat}_N(C^\infty)$. It can b done in two ways, either by letting

$$\left(\sqrt{3}\right)^{-1} > \alpha_1 > \alpha_2 > \cdots > \alpha_N > \left(2\sqrt{3}\right)^{-1}, \qquad C_j(1) = 0 \qquad (1 \leqslant j \leqslant N -$$
$$C_N(1) \geqslant 0, \qquad C_j(2) = e^{m_j}, \qquad C_j(3) = (-1)^{j-1} e^{-m_j},$$

or by setting

$$\begin{gathered}\left(\sqrt{3}\right)^{-1} > \alpha_1 > \alpha_2 > \cdots > \alpha_p > \left(2\sqrt{3}\right)^{-1},\\ \left(\sqrt{3}\right)^{-1} > -\tilde{\alpha}_1 > -\tilde{\alpha}_2 > \cdots > -\tilde{\alpha}_q > \alpha_p,\\ C_j(1) = 0 \quad (1 \leqslant j \leqslant p), \qquad \tilde{C}_j(1) = 0 \quad (1 \leqslant j \leqslant q),\\ C_j(2) = e^{m_j}, \qquad C_j(3) = (-1)^{j-1}e^{-m_j} \quad (1 \leqslant j \leqslant p),\\ \tilde{C}_j(2) = e^{\tilde{m}_j}, \qquad \tilde{C}_j(3) = (-1)^{j-1}e^{-\tilde{m}_j} \quad (1 \leqslant j \leqslant q),\\ \operatorname{Im} m_j = \operatorname{Im} \tilde{m}_k = 0, \qquad p + q = N.\end{gathered}$$

As a results, we obtain two systems of real functions

$$\begin{gathered}g_j = e^{\alpha_j x}\left(e^{\theta_j} + (-1)^{j-1}e^{-\theta_j}\right) \qquad (1 \leqslant j \leqslant N-1),\\ g_N = e^{-2\alpha_N x - 3(\alpha_N^2 - \beta_N^2)t}C_N(1) + e^{\alpha_N x}\left(e^{\theta_N} + (-1)^{N-1}e^{-\theta_N}\right)\end{gathered}$$

and

$$\begin{gathered}r_j = e^{\alpha_j x}\left(e^{\theta_j} + (-1)^{j-1}e^{-\theta_j}\right) \qquad (1 \leqslant j \leqslant p)\\ \tilde{r}_j = e^{\tilde{\alpha}_j x}(e^{\tilde{\theta}_j} + (-1)^{j-1}e^{-\tilde{\theta}_j}) \qquad (1 \leqslant j \leqslant q),\end{gathered}$$

where

$$\begin{gathered}\theta_j = -\beta_j x + 2\alpha_j\beta_j t + m_j, \qquad \tilde{\theta}_j = -\tilde{\beta}_j x + 2\tilde{\alpha}_j\tilde{\beta}_j t + \tilde{m}_j,\\ \beta_j = \sqrt{1 - 3\alpha_j^2}, \qquad \tilde{\beta}_j = \sqrt{1 - 3\tilde{\alpha}_j^2}.\end{gathered}$$

Each function satisfies the conditions of Corollary 2.2.1.

This is why the equation

$$-4v_{xx} + 3v_{tt} + \left(v_{xx} - 6v^2\right)_{xx} = 0$$

has the real solutions $v(x,t)$, $v(x,t) \in C^\infty$ of the form

$$v = -\frac{\partial^2}{\partial x^2}\ln \operatorname{Det} W, \qquad \tilde{v} = -\frac{\partial^2}{\partial x^2}\ln \operatorname{Det} \widetilde{W},$$

where

$$W = W\{\partial; g_1, \ldots, g_N\}, \qquad \widetilde{W} = W\{\partial; r_1, \ldots, r_p, \tilde{r}_1, \ldots, \tilde{r}_q\}.$$

These solutions depend on $2N+1$ and $2N$ real parameters, respectively.

Let now $\lambda = \varepsilon = 1$. In this case, the functions $f_j = f_j(x,t)$ can take complex values. Since the functions $\overline{f_j} = \overline{f_j(x,t)}$ also satisfy Eqs. (2.3.7), if we replace in them $\overline{d_j}, \bar{a}_j$ for d_j, a_j, then the Wronsky matrix

$$W = W\{\partial; f_1, \ldots, f_N, \bar{f}_1, \ldots, \bar{f}_N\} \in \operatorname{Mat}_{2N}(C^\infty)$$

satisfies Eqs. (2.3.4) when

$$D = \operatorname{diag}(d_1, \dots, d_N, \bar{d}_1, \dots, \bar{d}_N), \qquad A = \operatorname{diag}(a_1, \dots, a_N, \bar{a}_1, \dots, \bar{a}_N),$$

while $\overline{\operatorname{Det} W} = (-1)^N \operatorname{Det} W$.

Therefore, in the domain where the variables x, t change and where the matrix is invertible, (2.3.5) yields an infinitely differentiable solution of the equation

$$4v_{xx} + 3v_{tt} + \left(v_{xx} - 6v^2\right)_{xx} = 0.$$

We obtain the simplest one at $N = 1$, if we set $C(1) = 0$ in formula (2.3.9):

$$v = -\frac{\partial^2}{\partial x^2} \ln\left[\operatorname{ch}(4\alpha\beta t + m) + \frac{\alpha}{\sqrt{4\alpha^2 - 1}} \cos(2\beta x + \varphi)\right],$$

$$\frac{1}{2} < \alpha < \frac{1}{\sqrt{3}}, \qquad \beta = \sqrt{1 - 3\alpha^2}, \qquad \operatorname{Im} m = \operatorname{Im} \varphi = 0.$$

2 *Toda and Langmuir Lattices*

Solutions of these equations are sought in the ring $K_0 = C^\infty(Z \times R^1)$ of scalar functions where a differentiation ∂, an automorphism α, and g.d. ∂_α are defined by the formulas

$$\partial f = \frac{\partial f}{\partial t}, \qquad \alpha(f(n,t)) = f(n+1,t), \qquad \partial_\alpha(f(n,t)) = f(n+1,t) - f(n,t).$$

Formulas II(6), II(8), together with the equalities (2.1.8) and (2.1.9), imply that if the invertible Wronsky matrices

$$W_1 = W\{\partial; f_1, \dots, f_N\} \in \operatorname{Mat}_N(C^\infty(Z \times R^1)),$$
$$W_2 = W\{\partial; g_1, \dots, g_N\} \in \operatorname{Mat}_N(C^\infty(Z \times R^1))$$

satisfy the equations

(2.3.10) $$W_l(n+2,t) + W_l(n,t) = AW_l(n+1,t),$$

(2.3.10′) $$\partial W_l(n,t) = W_l(n+l,t) \qquad (l = 1,2),$$

where $A \in \operatorname{Mat}_N$ is a constant matrix, then the functions

(2.3.11) $$u_1(n) = u_1(n,t) = (-1)^{N-1} \frac{\operatorname{Det} W_1(n+1,t)}{\operatorname{Det} W_1(n,t)}$$

(2.3.12) $$u_2(n) = u_2(n,t) = \frac{\operatorname{Det} W_2(n-1,t) \operatorname{Det} W_2(n+2,t)}{\operatorname{Det} W_2(n,t) \operatorname{Det} W_2(n+1,t)}$$

satisfy the equations

(2.3.13) $$\partial(u_1(n)^{-1} \partial u_1(n)) = u_1(n)^{-1} u_1(n+1) - u_1(n) u_1(n-1)^{-1}$$

(2.3.14) $$\partial u_2(n) = u_2(n)[u_2(n+1) - u_2(n-1)].$$

If $A = \text{diag}(a_1, \ldots, a_N)$, then Eq. (2.3.10) is split into finite difference equations in f_j, g_j:

$$y(n+2) + y(n) = ay(n+1) \qquad (a = a_1, \ldots, a_N)$$

whose general solution is given by the formula

$$y(n) = \beta^n C_1(t) + \beta^{-n} C_2(t) \qquad (\beta + \beta^{-1} = a).$$

Substituting this expression into Eq. (2.3.10), we find that the functions f_j, g_j have the form:

$$f = \lambda^n e^{\lambda t} C_1 + \lambda^{-n} e^{\lambda^{-1} t} C_2,$$
$$g = \mu^n e^{\mu^2 t} D_1 + \mu^{-n} e^{\mu^{-2} t} D_2,$$

where λ, μ, C, D are arbitrary numerical parameters.

By choosing the parameters λ, μ in such a way that the inequalities

$$-1 < \lambda_1 < \cdots < \lambda_p < 0 < \tilde{\lambda}_q < \cdots < \tilde{\lambda}_1 < 1 \qquad (p + q = N),$$
$$0 < \mu_N^2 < \cdots < \mu_1^2 < 1$$

will hold, we shall see that the functions

$$\text{(2.3.15)} \qquad \begin{aligned} f_j &= \lambda_j^n e^{\lambda_j t + m_j} + (-1)^{j-1} \lambda_j^{-n} e^{\lambda_j^{-1} t - m_j} \qquad (1 \leqslant j \leqslant p), \\ \tilde{f}_k &= \tilde{\lambda}_k^n e^{\tilde{\lambda}_k t + \tilde{m}_k} + (-1)^{k-1} \tilde{\lambda}_k^{-n} e^{\tilde{\lambda}_k t - \tilde{m}_k} \qquad (1 \leqslant k \leqslant q), \end{aligned}$$

and

$$\text{(2.3.16)} \qquad g_j = \mu_j^n e^{\mu_j^2 t + m_j} + (-1)^{j-1} \mu_j^{-n} e^{\mu_j^{-2} t - m_j}$$

satisfy the conditions of Corollary 2.2.1 if the numbers $m_j, \tilde{m}_j$ are real. Hence the matrices

$$W_1 = W\{\partial; f_1, \ldots, f_p, \tilde{f}_1, \ldots, \tilde{f}_q\}$$

and

$$W_2 = W\{\partial; g_1, \ldots, g_N\}$$

are invertible in the ring $\text{Mat}_N(C^\infty(Z \times R^1))$, and the functions (2.3.11) and (2.3.12) belong to the ring $C^\infty(Z \times R^1)$ and satisfy Eqs. (2.3.13) and (2.3.14).

Besides, Eqs. (2.3.15) and (2.3.16) imply that

$$\lim_{t \to \pm\infty} = \frac{\text{Det}\, W_1(n+1, t)}{\text{Det}\, W_1(n, t)} = (\lambda_1 \ldots \lambda_p \tilde{\lambda}_1^{-1} \ldots \tilde{\lambda}_q^{-1})^{\pm 1},$$
$$\lim_{t \to \pm\infty} = \frac{\text{Det}\, W_2(n+1, t)}{\text{Det}\, W_2(n, t)} = (\mu_1 \ldots \mu_N)^{\mp 1},$$

and since the functions $u_1(n,t)$, $u_2(n,t)$ are real and nonzero for all $t \in (-\infty, \infty)$,

$$(-1)^{N+p-1} u_1(n,t) > 0, \qquad u_2(n,t) > 0.$$

Hence, the functions

$$v_1(n) = v_1(n,t) = \ln[(-1)^{N+p-1}u_1(n,t)],$$
$$v_2(n) = v_2(n,t) = \ln u_2,(n,t),$$

are real, belong to the ring $C^\infty(Z \times R^1)$ and satisfy the equation

$$\frac{d^2}{dt^2}v_1(n) = e^{v_1(n+1)-v_1(n)} - e^{v_1(n)-v_1(n-1)},$$
$$\frac{d}{dt}v_2(n) = e^{v_2(n+1)} - e^{v_2(n-1)}.$$

3 *Sine–Gordon equation*

Define the differentiations ∂, ∂_1 and the automorphism β for scalar functions $f(x,y)$ in the ring $C^\infty(R^2)$ by the equations

$$\partial_f = \frac{\partial f}{\partial x}, \qquad \partial_1 f = \frac{\partial f}{\partial y}, \qquad \beta(f(x,y)) = \overline{f(x,y)}.$$

Formula II(7′) and the equalities (2.1.8), (2.1.9) imply that if a matrix

$$W = W\{\partial; f_1,\dots,f_N\} \in \mathrm{Mat}_N(C^\infty(R^2))$$

is invertible and satisfies the equations

$$\partial_1(\partial W) = W, \qquad \partial W = A\overline{W}, \tag{2.3.17}$$

where $A \in \mathrm{Mat}_N$ is a constant matrix, then the function

$$\gamma_N = \gamma_N(x,y) = (-1)^{N-1}(\mathrm{Det}\, W)^{-1}\,\mathrm{Det}(\partial W)$$

satisfies the equation

$$\partial_1(\gamma_N^{-1}\partial\gamma_N) = \bar{\gamma}_N^{-1}\gamma_N - \gamma_N^{-1}\bar{\gamma}_N. \tag{2.3.18}$$

The Wronsky matrices for which

$$f_j = e^{\theta_j} + i(-1)^{n_j}e^{-\theta_j}, \qquad \theta_j = \lambda_j x + \lambda_j^{-1}y + r(\lambda_j)$$

(n_j are arbitrary integers) obviously satisfy the first of the Eqs. (2.3.17); if they also satisfy the conditions of Corollary 2.2.2, then they solve the second equation with $A = \Lambda M$, are invertible, and

$$\gamma_N = (-1)^{N-1+r}\Big(\prod_{j=1}^{N}\lambda_j\Big)e^{-2i\varphi}, \qquad \varphi = \varphi(x,y) = \arg \mathrm{Det}\, W. \tag{2.3.19}$$

To meet the conditions of Corollary 2.2.2, one must take an arbitrary symmetric set from the open right half-plane in the capacity of $\lambda_1,\dots,\lambda_N$ and fulfill the condition $\overline{r(\lambda_j)} = r(\bar{\lambda}_j)$. After that, the pre-exponent factor in Eq. (2.3.19) becomes a real number, and Eq. (2.3.18) is reduced to

$$2i\partial_1\partial\varphi = \exp(-4i\varphi) - \exp(4i\varphi).$$

From this, we find, after elementary transformations, that the functions

$$u = u(x, y) = 4 \arg \operatorname{Det} W \in C^\infty(R^2)$$

are real solutions of the equation

$$\frac{\partial^2 u}{\partial x \partial y} = 4\sin(u + \pi N)$$

if

$$W = W\{\partial; g_1, \ldots, g_N\},$$

where

$$g_j = \operatorname{ch}(\lambda_j x + \lambda_j^{-1} y + r(\lambda_j) - \frac{i\pi}{4}(2n_j + 1)).$$

Here $\lambda_1, \ldots, \lambda_N$ is an arbitrary symmetrical set from the open right half-plane, $\overline{r(\lambda_j)} = r(\bar{\lambda}_j)$ are arbitrary complex numbers, and n_j are arbitrary integers.

4 *Schrödinger and Heisenberg Nonlinear Equations*

The solutions of these equations one finds in the ring $K_0 = \operatorname{Mat}_r(C^\infty(R^2))$ of matrix functions $F(x, t)$ of the order $r (r \geqslant 2)$ with differentiations

$$\partial_0 F = i\frac{\partial F}{\partial t}, \qquad \partial F = \frac{\partial F}{\partial x}.$$

We shall denote the unit of this ring by I, by B the matrix

$$B = \operatorname{diag}(\underbrace{1, \ldots, 1}_{r_1}, \underbrace{-1, \ldots, -1}_{r_2}) \in \operatorname{Mat}_r \qquad (r = r_1 + r_2),$$

and by Q_1, Q_2 the orthoprojectors

$$Q_1 = \frac{1}{2}(I + B), \qquad Q_2 = \frac{1}{2}(I - B).$$

As an extension of K_0 we shall take the ring $K = \operatorname{Mat}_N(K_0)$ of block matrices with elements from K_0 and shall set

$$I_N = \operatorname{diag}(I, \ldots, I), \qquad B_N = \operatorname{diag}(B, \ldots, B), \qquad P = \operatorname{diag}(I, 0, \ldots, 0).$$

It is obvious that I_N is a unity in K, $B_N^2 = I_N$, and $[B_N, P] = 0$.

Formulas II(3′, 5) imply that if an invertible Wronsky matrix

$$W = W\{\partial; f_1, \ldots, f_N\} \in \operatorname{Mat}_N(K_0)$$

satisfies the equations

$$i\frac{\partial u}{\partial t} + \frac{\partial^2 W}{\partial x^2} B_N = 0, \qquad \frac{\partial W}{\partial x} = AWB_N, \tag{2.3.20}$$

where $A \in \mathrm{Mat}_N(\mathrm{Mat}_r)$ is a constant matrix, then the matrices

$$u = [P\gamma P, B_N], \qquad S = PN^{-1}\gamma B_N \gamma^{-1} NP \qquad (\gamma = W^{-1}\partial W)$$

belong to the ring K_0 and satisfy the equations

$$2i\frac{\partial u}{\partial t} B + \frac{\partial^2 u}{\partial x^2} + 2u^3 = 0, \tag{2.3.21}$$

$$4i\frac{\partial S}{\partial t} = [S, \frac{\partial^2 S}{\partial x^2}]. \tag{2.3.22}$$

Besides, according to (2.1.6), (2.1.8),

$$u = [\gamma_1, B], \qquad S = \gamma_N B \gamma_N^{-1},$$

where the matrices $\gamma_1, \gamma_N \in \mathrm{Mat}_r(C^\infty)$ are found from the system of equations (2.1.7).

Let us take in the capacity of A a diagonal matrix $A = \mathrm{diag}(a_1, \ldots, a_N)(a_i \in \mathrm{Mat}_r$ is a constant). Then Eqs. (2.3.20) will split into such a system of equations in the matrices $f_j = f_j(x,t) \in \mathrm{Mat}_r(C^\infty)$:

$$i\frac{\partial f_j}{\partial t} + \frac{\partial^2 f_j}{\partial x^2} B = 0, \qquad \frac{\partial f_j}{\partial x} = a_j f_j B. \tag{2.3.20$'$}$$

From these equations and the equalities $BQ_1 = Q_1 B = Q_1$, $BQ_2 = Q_2 B = -Q_2$, it follows that

$$\begin{aligned} &i\frac{\partial f_j}{\partial t} Q_p + (-1)^{p+1} \frac{\partial^2 f_j}{\partial x^2} Q_p = 0, \\ &\frac{\partial f_j}{\partial x} Q_p = (-1)^{p+1} a_j f_j Q_p \qquad (p = 1, 2). \end{aligned}$$

Hence

$$f_j Q_p = \exp\{(-1)^{p+1}(a_j x + i a_j^2 t)\} C_j(p) Q_p,$$

and

$$f_j = f_j(Q_1 + Q_2) = e^{a_j x + i a_j^2 t} C_j(1) Q_1 + e^{-(a_j x + i a_j^2 t)} C_j(2) Q_2,$$

where $C_j(1), C_j(2) \in \mathrm{Mat}_r$ are arbitrary constant matrices.

Let us choose the parameters $a_j, C_j(1), C_j(2)$ so that they would meet the conditions of the Lemma 2.2.3 which will guarantee invertibility of the matrix $W\{\partial; f_1, \ldots, f_N\}$. For this, we must set

$$a_j = \lambda_j Q_1 - \bar{\lambda}_j Q_2, \qquad C_j(1) = e^{\alpha_j} Q_1 - e^{-\bar{\alpha}_j} Q_2 C_j^* Q_1,$$
$$C_j(2) = e^{\bar{\alpha}_j} Q_2 + e^{-\alpha_j} Q_1 C_j Q_2,$$

where $\alpha_j, \lambda_j(\operatorname{Re}\lambda_j > 0)$ are arbitrary complex numbers, and C_j are arbitrary constant matrices of the order r.

Then the matrices f_j will have the form

$$f_j = e^{\theta_j}Q_1 + e^{-\theta_j}D_j - e^{-\bar\theta_j}D_j^* + e^{\bar\theta_j}Q_2, \tag{2.3.23}$$
$$\theta_j = \lambda_j x + i\lambda_j^2 t + \alpha_j, \qquad D_j = Q_1 C_j Q_2.$$

Thus, Wronsky matrices $W\{\partial; f_1, \ldots, f_N\}$ with the functions $f_j = f_j(x,t) \in \operatorname{Mat}_r(C^\infty)$ defined by Eq. (2.3.23) are invertible and satisfy Eq. (2.3.20′), while matrices

$$u(x,t) = [\gamma_1, B], \qquad S = \gamma_N B \gamma_N^{-1}$$

satisfy Eqs. (2.3.21) and (2.3.22). Besides, by virtue of Corollary 2.2.3, the matrix u is Hermitian ($u^* = u$), and the matrix $\tilde\gamma_N = (\prod_{j=1}^N |\lambda_j|)^{-1}\gamma_N$ is unitary ($\tilde\gamma_N^{-1} = \tilde\gamma_N^*$). That is why the matrix

$$S = \gamma_N B \gamma_N^{-1} = \tilde\gamma_N B \tilde\gamma_N^*$$

is both unitary and Hermitian, while

$$u = \begin{pmatrix} 0 & v \\ v^* & 0 \end{pmatrix},$$

where $v = v(x,t)$ is a rectangular matrix function possessing r_1 rows and r_2 columns. Substituting this expression into Eq. (2.3.21), we find that the matrix v satisfies the equation

$$2i\frac{\partial v}{\partial t} = \frac{\partial^2 v}{\partial x^2} + 2vv^*v,$$

which is equivalent to the system of equations

$$2i\frac{\partial \vec y_p}{\partial t} = \frac{\partial^2 \vec y_p}{\partial x^2} + 2\sum_{j=1}^{r_1}(\vec y_p, \vec y_j)\vec y_j \qquad (1 \leqslant p \leqslant r_1)$$

in r_2-dimensional vector functions $\vec y_p = \vec y_p(x,t) = (y_{p1}, y_{p2}, \ldots, y_{pr_2})$, where y_{pj} are elements of the matrix v.

We present the final expressions for one-soliton solutions ($N = 1$). In this case,

$$W^{-1}\partial W = \gamma = \begin{pmatrix} \gamma_{11} & \gamma_{12} \\ \gamma_{21} & \gamma_{22} \end{pmatrix},$$

where

$$\gamma_{11} = \lambda A_1^{-1} - \bar\lambda e^{-4R} D A_2^{-1} D^*, \qquad \gamma_{22} = \bar\lambda A_2^{-1} - \lambda e^{-4R} D^* A_1^{-1} D,$$
$$\gamma_{12} = -\gamma_{21}^* = -e^{-2\theta}\{\lambda A_1^{-1} D + \bar\lambda D A_2^{-1}\}.$$

Here D is an arbitrary constant matrix with r_1 rows and r_2 columns, $\theta = \lambda x + i\lambda^2 t + \alpha$, $\lambda(\operatorname{Re}\lambda > 0)$ and α are arbitrary complex numbers, $R = \operatorname{Re}\theta$,

$$A_1, = I_1 + e^{-4R} DD^*, \qquad A_2 = I_2 + e^{-4R} D^* D,$$

and I_1, I_2 are unit matrices of orders r_1, and r_2, respectively.

One-soliton solutions of Eqs. (2.3.21) and (2.3.22) have the form:

$$v = -2\gamma_{12} = 2e^{-2\theta}\{\lambda A_1^{-1}D + \bar{\lambda} D A_2^{-1}\},$$
$$S = |\lambda|^{-2}\begin{pmatrix} 2\gamma_{11}\gamma_{11}^* - |\lambda|^2 I_1 & -2\gamma_{11}\gamma_{12} \\ -2\gamma_{12}^*\gamma_{11}^* & -2\gamma_{22}\gamma_{22}^* + |\lambda|^2 I_2 \end{pmatrix}.$$

In particular, if $r_1 = r_2 = 1$, then

$$v = (\lambda + \bar{\lambda})\frac{\exp(-2iI)}{\operatorname{ch} 2R},$$
$$S = \begin{pmatrix} s_{11} & s_{12} \\ \bar{s}_{12} & -s_{11} \end{pmatrix},$$
$$s_{11} = 1 - \frac{2\cos^2\varphi}{\operatorname{ch}^2 2R},$$
$$s_{12} = \frac{2\cos\varphi \operatorname{sh}(2R + i\varphi)\exp(-2iI)}{\operatorname{ch}^2 2R}$$

where

$$\lambda = |\lambda| e^{i\varphi}, \qquad R = \operatorname{Re}\theta, I = \operatorname{Im}\theta$$

and

$$\theta = \lambda x + i\lambda^2 t + \alpha.$$

5 *Kadomtsev–Petviashvili Equations and Generalized Toda Lattice Equations*

Kadomtsev–Petviashvili (KP) equations are considered in the ring $K_0 = C^\infty(R_3)$ of scalar functions $f(x, y, t)$ with differentiations

$$\partial f = \frac{\partial f}{\partial x}, \qquad \partial_1 f = \frac{\partial f}{\partial t}, \qquad \partial_2 f = \frac{\partial f}{\partial y}.$$

Results of §1 of this chapter and formulas II(1) imply that if a matrix $W = W\{\partial; f_1; \ldots; f_N\} \in \operatorname{Mat}_N(C^\infty)$ is invertible and satisfies the equations

$$\partial^3 W + \partial_1 W = AW, \qquad \partial^2 W + \varepsilon^{-1}\partial_2 W = BW \tag{2.3.24}$$

($A, B \in \operatorname{Mat}_N$ are constant matrices), then the function

$$v = v(x, y, t) = -\frac{\partial^2}{\partial x^2}, \ln \operatorname{Det} W \tag{2.3.25}$$

belongs to the ring $C^\infty(R^3)$ and satisfies the KP equation

$$\frac{\partial}{\partial x}\left\{4\frac{\partial v}{\partial t} + \frac{\partial^3 v}{\partial x^3} - 12v\frac{\partial v}{\partial x}\right\} + 3\varepsilon^2\frac{\partial^2 v}{\partial y^2} = 0. \tag{2.3.26}$$

Without loss of generality we can restrict ourselves to the case when $A = B = 0$ and Eqs. (2.3.24) are reduced to

$$\partial^3 f_j + \partial_1 f_j = 0, \qquad \partial^2 f_j + \varepsilon^{-1}\partial_2 f_j = 0 \qquad (1 \leqslant j \leqslant N).$$

The functions

$$f_j = \int e^{\xi x - \xi^3 t - \varepsilon \xi^2 y} C_j(\xi) d\sigma(\xi)$$

evidently belong to the ring $C^\infty(R^3)$ and solve these equations if the nonnegative measure $d\sigma(\xi)$ satisfies the inequalities

$$\int e^{|\xi^3 t|} |C_j(\xi)| d\sigma(\xi) < \infty \qquad (t \in R^1, \quad 1 \leqslant j \leqslant N).$$

Since we are interested only in real solutions of the KP equation, functional parameters $C_j(\xi), d\sigma(\xi)$ must be chosen so that $\operatorname{Det} W$ would be real and would not vanish for all real values of x, y, t. If $\varepsilon = 1$, then the parameters can be selected, making use of Corollary 2.2.1. For this purpose we shall take a finite system of disjoint half-intervals $\Delta_1, \Delta_2, \ldots, \Delta_l$ and their partitions

$$\{\alpha_j(\Delta_k), \beta_j(\Delta_k)\} \qquad (1 \leqslant j \leqslant n_k, \sum n_k = N).$$

If the measure $d\sigma(\xi)$ is concentrated on the set $M = \cup_{k=1}^l \Delta_k$, and

$$\int_M e^{|\xi^3 t|} d\sigma(\xi) < \infty, \qquad \int_{\alpha_j(\Delta_k)} d\sigma(\xi) + \int_{\beta_j(\Delta_k)} d\sigma(\xi) > 0,$$

then the functions

$$f_j^{\Delta_k} = \int_{\alpha_j(\Delta_k)} e^{\xi x - \xi^3 t - \xi^2 y} d\sigma(\xi) + (-1)^{j-1} \int_{\beta_j(\Delta_k)} e^{\xi x - \xi^3 t - \xi^2 y} d\sigma(\xi)$$

$(1 \leqslant j \leqslant n_k,\ 1 \leqslant k \leqslant l)$ satisfy the conditions of Corollary 2.2.1. Therefore, the matrix $W = W\{\partial; f_1^{\Delta_1}, \ldots, f_{n_1}^{\Delta_1} f_1^{\Delta_l}, \ldots, f_{n_l}^{\Delta_l}\}$ belongs to the ring $\operatorname{Mat}_N(C^\infty)$ and is invertible in it, while the function (2.3.25) is a regular real solution of the KP equation with $\varepsilon = 1$.

Solutions of the generalized Toda lattice equation

$$\frac{\partial^2 y(n)}{\partial x \partial y} = e^{y(n) - y(n-1)} - e^{y(n+1) - y(n)} \tag{2.3.27}$$

are found likewise, and we present only the final formulas. If the nonnegative measure $d\sigma(\xi)$ is concentrated on the set

$$M = \bigcup_{k=1}^{l} \Delta k, \qquad 0 \notin M,$$

and for all $x \in R^1$, $y \in R^1$,

$$\int_M e^{\xi x + \xi^{-1} y} d\sigma(\xi) < \infty, \qquad \int_{\alpha_j(\Delta_k)} d\sigma(\xi) + \int_{\beta_j(\Delta_k)} d\sigma(\xi) > 0,$$

then the functions

$$f_j^{\Delta_k} = \int\limits_{\alpha_j(\Delta_k)} e^{\xi x + \xi^{-1} y} \xi^n d\sigma(\xi) + (-1)^{j-1} \int\limits_{\beta_j(\Delta_k)} e^{\xi x + \xi^{-1} y} \xi^n d\sigma(\xi)$$

$(1 \leqslant j \leqslant n_k, 1 \leqslant k \leqslant l, \sum n_k = N)$ satisfy the conditions of Corollary 2.2.1, the matrix

$$W = W(n, x, y) = W\left\{\partial; f_1^{\Delta_1}, \ldots, f_{n_1}^{\Delta_1}, \ldots, f_1^{\Delta_l}, \ldots, f_{n_l}^{\Delta_l}\right\}$$

is invertible, and the functions

$$y(n) = y(n, x, y) = \ln \frac{\operatorname{Det} W(n+1, x, y)}{\operatorname{Det} W(n, x, y)} \tag{2.3.28}$$

satisfy Eqs. (2.3.27).

Note that the obtained solutions of Eqs. (2.3.26) and (2.3.27) depend on the parameter $d\sigma(\xi)$. Soliton solutions are obtained when the measure $d\sigma(\xi)$ is discrete. If it is concentrated in the points $0 < p_N < p_{N-1} < \cdots < p_1 < q_1 < q_2 < \cdots < q_N$, then

$$f_j = e^{p_j x - p_j^3 t - p_j^2 y + a_j} + (-1)^{j-1} e^{q_j x - q_j^3 t - q_j^2 y + b_j}$$

and, accordingly,

$$f_j = p_j^n e^{p_j x + p_j^{-1} y + a_j} + (-1)^{j-1} q_j^n e^{q_j x + q_j^{-1} y + b_j}$$

(a_j, b_j are arbitrary real numbers, while the functions (2.3.25), and (2.3.28) are N-soliton solutions of Eqs. (2.3.26) with $\varepsilon = 1$ and (2.3.27)).

6 *Generalized N-Wave Problem*

The equation

$$\partial_2[u, D_1] - \partial_1[u, D_2] - \partial([u, D_1]D_2 - [u, D_2]D_1) = [[u, D_1], [u, D_2]]$$

is considered in the ring $K_0 = \mathrm{Mat}_r(C^\infty(K^3))$ of the matrix functions $u = u(x, x_1, x_2)$ of the order r with differentiations

$$\partial u = \frac{\partial u}{\partial x}, \qquad \partial_1 u = \frac{\partial u}{\partial x_1}, \qquad \partial_2 u = \frac{\partial u}{\partial x_2}.$$

Let us denote diagonal matrices of the order r by

$$d = d(z, x), \qquad d_1 = d_1(x_1), \qquad d_2 = d_2(x_2) \in \mathrm{Mat}_r(C^\infty),$$

their derivatives by

$$\begin{aligned} D &= D(z, x) = \partial(d(z, x)), \\ D_1 &= D_1(x_1) = \partial_1(d_1(x_1)), \\ D_2 &= D_2(x_2) = \partial_2(d_2(x_2)), \end{aligned}$$

and the arbitrary orthoprojectors by $Q_1(z), Q_2(z) = I - Q_1(z) \in \mathrm{Mat}_r$. We shall choose these matrices to meet the following conditions: 1) $d_1 = d_1^*, d_2 = d_2^*$; 2) the matrices D_1, D_2 are invertible; 3) the orthoprojectors $Q_1(z)$, $Q_2(z)$ are commutative with the matrix $D(z,x)$.

Since the matrices

$$\mathcal{E}(z) = \exp\{d(z,x) + z(d_1(x_1) + d_2(x_2))\},$$
$$\mathcal{E}(z)^{*-1} = \exp\{-d(z,x)^* - \bar{z}(d_1(x_1) + d_2(x_2))\}$$

satisfy the equations

$$\partial_\alpha \mathcal{E}(z) = \{\partial \mathcal{E}(z) + (zI - D(z,x))\mathcal{E}(z)\}D_\alpha,$$
$$\partial_\alpha \mathcal{E}(z)^{*-1} = \{\partial \mathcal{E}(z)^{*-1} - (\bar{z}I - D(z,x)^*)\mathcal{E}(z)^{*-1}\}D_\alpha, \qquad (\alpha = 1,2)$$

and the projectors $Q_1(z)$, $Q_2(z)$ are orthogonal and commutative with $D(z,x)$, the matrices

$$\Gamma_j = \Gamma_j(z) = Q_1(z)z^{N-j}\mathcal{E}(z) + Q_2(z)(-z)^{N-j}\mathcal{E}(z)^{*-1}$$

satisfy the equations

$$\partial_\alpha \Gamma_j = \{\partial \Gamma_j + A(z,x)\Gamma_j\}D_\alpha \qquad (\alpha = 1,2)$$

where

$$A(z,x) = Q_1(z)(zI - D(z,x)) - Q_2(z)(\bar{z}I - D(z,x)^*).$$

Hence, the block matrix $\Gamma = (\Gamma_{ij}) \in \mathrm{Mat}_N(\mathrm{Mat}_r(C^\infty(R^3)))$, with elements

$$\Gamma_{ij} = \Gamma_j(z_i) \qquad (i,j = 1,2,\ldots,N),$$

satisfies the equations

$$\partial_\alpha \Gamma = (\partial\Gamma + A(x)\Gamma)\tilde{D}_\alpha \qquad (\alpha = 1,2),$$

where

$$A(x) = \mathrm{diag}\{A(z_1,x), \qquad A(z_2,x),\ldots,A(z_N,x)\},$$
$$\tilde{D}_\alpha = \mathrm{diag}\{D_\alpha, D_\alpha, \ldots, D_\alpha\}$$

are diagonal block matrices; the matrices $\tilde{D}_\alpha$are invertible and commutative, with the matrix P defined by equality (2.1.1). Further, the equalities $\partial_\alpha \Gamma_j(z_i) = \Gamma_{j-1}(z_i)D_\alpha$ (as in the case of the Wronsky matrices) imply that $\partial_\alpha \Gamma(I-P) = \Gamma N \tilde{D}_\alpha(I-P)$, while from Lemma 2.2.4 it follows that the matrix Γ is invertible if the numbers $z_1, z_2, \ldots, z_N$ are different and lie in the right open half-plane. Thus, all the conditions under which formulas II(9) were obtained are met, and the matrix

$$u_1 = P\Gamma^{-1}\partial_1\Gamma\tilde{D}_1^{-1}P = P\Gamma^{-1}\partial_2\Gamma\tilde{D}_2^{-1}P$$

satisfies Eq. (2.3.29). Since this solution is found from the system of linear equations

$$\sum_{j=1}^{N} \Gamma_j(z_i)u_j = \Gamma_0(z_i) \qquad (1 \leqslant i \leqslant N),$$

then, according to Corollary 2.2.4, the matrix $u_1 + i\alpha I = (u_1 + i\alpha I)^*$ is a self-adjoint solution of Eq. (2.3.29) in the ring $K_0 = \mathrm{Mat}_r(C^\infty(R^3))$.

The final formula for one-soliton self-adjoint solutions of this equation has such a form

$$u = \frac{z+\bar{z}}{2}(Q_1\mathcal{E}(z) + Q_2\mathcal{E}(z)^{*-1})^{-1}(Q_1\mathcal{E}(z) - Q_2\mathcal{E}(z)^{*-1},$$

where Q_1 and $Q_2 = I - Q_1$ are arbitrary orthoprojectors, $z(\mathrm{Re}\, z > 0)$ is an arbitrary complex number, and

$$\mathcal{E}(z) = \exp\{d(x) + z(d_1(x_1) + d_2(x_2))\}.$$

7 *Chiral Field Equation*

Solutions of the equation

$$2\partial_1\partial_2(S) = (\partial_1 S)S^{-1}(\partial_2 S) + (\partial_2 S)S^{-1}(\partial_1 S) \tag{2.3.30}$$

are sought in the ring $K_0 = \mathrm{Mat}_r(C^\infty(R^2))$ with differentiations

$$\partial_1 S = \frac{\partial S}{\partial x_1}, \qquad \partial_2 S = \frac{\partial S}{\partial x_2}.$$

Find at first one-soliton solutions of this equation. Let diagonal matrices

$$d_1 = d_1(x_1), \qquad d_2 = d_2(x_2) \in \mathrm{Mat}_r(C^\infty(R^2))$$

and their derivatives

$$D_1 = D_1(x_1) = \partial_1(d_1(x_1)), \qquad D_2 = D_2(x_2) = \partial_2(d_2(x_2))$$

satisfy the conditions 1) $d_1 = d_1^*, d_2 = d_2^*$; 2) matrices D_1, D_2 are invertible. The matrix $\mathcal{E}(w_1, w_2) = \exp\{w_1 d_1(x_1) + w_2 d_2(x_2)\}$ obviously satisfies Eq. (1.2.11) with the coefficients $A_1 = w_1 I$, $A_2 = w_2 I$. In order to meet the relation (1.2.11) one must set $w_1 = 2\lambda_2(1-iz)^{-1}$, $w_2 = 2\lambda_1(1+iz)^{-1}$, where z is an arbitrary complex parameter. In particular, if $2\lambda_1 = 2\lambda_2 = i$, then $w_1 = i(1-iz)^{-1}, w_2 = i(1+iz)^{-1}$. Thus, the matrices

$$\mathcal{E}(z) = \exp i\left\{(1-iz)^{-1}d_1(x_1) + (1+iz)^{-1}d_2(x_2)\right\}$$

solve Eqs. (1.2.11) with coefficients

$$A_1 = w_1(z)I = i(1-iz)^{-1}I, \qquad A_2 = w_2(z)I = i(1+iz)^{-1}I$$

that satisfy the identity $2w_1(z)w_2(z) \equiv i(w_1(z)+w_2(z))$. Having selected arbitrary orthoprojectors $Q_1(z), Q_2(z) = I - Q_1(z) \in \mathrm{Mat}_r$, we obtain the matrix

$$\Gamma = Q_1(z)\mathcal{E}(z) + Q_2(z)\mathcal{E}(-\bar{z})$$

satisfying the equations

$$\partial_\alpha \Gamma = A_\alpha(z)\Gamma D_\alpha \qquad (\alpha = 1,2)$$

with the coefficients

$$A_\alpha(z) = Q_1(z)w_\alpha(z) + Q_2(z)w_\alpha(-\bar{z})$$

related by the equality $2A_1(z)A_2(z) = i(A_1(z)+A_2(z))$. The equality $\mathcal{E}(-\bar{z}) = \mathcal{E}(z)^{*-1}$ and Lemma 2.2.4 imply that the matrix Γ is invertible. Hence, the matrix

$$v = \Gamma^{-1}(I - iA_1(z)^{-1}\Gamma = i\Gamma^{-1}(Q_1(z)z - Q_2(z)\bar{z})\Gamma$$

belongs to the ring $\mathrm{Mat}_r(C^\infty(R^2))$ and, according to corollary 2.2.4, $v^*v = z\bar{z}I$. Note, finally, that the matrix $\mathcal{E}(0) = \exp i(d_1(x_1)+d_2(x_2))$ satisfies the equations $\partial_\alpha(\mathcal{E}(0)) = i\mathcal{E}(0)D_\alpha$ whence, resorting to formulas II(10), we find that the matrix

$$S = \mathcal{E}(0)\Gamma^{-1}(Q_1(z)\frac{z}{|z|} - Q_2(z)\frac{\bar{z}}{|z|})\Gamma$$

is a unitary solution of Eq. (2.3.30).

In order to find multisoliton solutions of this equation, one must consider the matrices

$$\Gamma_j(z) = Q_1(z)z^{N-j}\mathcal{E}(z) + Q_2(z)(-\bar{z})^{N-j}\mathcal{E}(-\bar{z})$$

and build with them the block matrix

$$\Gamma = (\Gamma_{ij}) \in \mathrm{Mat}_N(\mathrm{Mat}_r(C^\infty(R^2)))$$

with the elements $\Gamma_{ij} = \Gamma_j(z_i)$. Here $z_1, z_2, \ldots, z_N$ is an arbitrary set of different complex numbers from the open right plane, and

$$Q_1(z), \qquad Q_2(z) = I - Q_1(z) \in \mathrm{Mat}_r$$

are arbitrary orthoprojectors. The matrix we have constructed satisfies the equation

$$\partial_\alpha \Gamma = A_\alpha \Gamma \tilde{D}_\alpha \qquad (\alpha = 1,2),$$

where

$$A_\alpha = \mathrm{diag}\{A_\alpha(z_1), A_\alpha(z_2), \ldots, A_\alpha(z_N)\},$$
$$\tilde{D}_\alpha = \mathrm{diag}\{D_\alpha, D_\alpha, \ldots, D_\alpha\}$$

are diagonal block matrices; besides, $2A_1A_2 = i(A_1 + A_2)$ and the matrices $\widetilde{D}_\alpha$ are invertible and commutative with the matrix P. Since

$$C = I - iA_1^{-1} = i\,\mathrm{diag}\{Q_1(z_1)z_1 - Q_2(z_2)\bar{z}_2, \ldots, Q_1(z_N)z_N - Q_2(z_N)\bar{z}_N\}$$

and, according to Lemma 2.2.4, the matrix Γ is invertible, the matrix $v = \Gamma^{-1}(-iC)\Gamma$ belongs to the ring $\mathrm{Mat}_N(\mathrm{Mat}_r(C^\infty(R^2)))$, while the matrix $\widetilde{\mathcal{E}}(0)v$ with

$$\widetilde{\mathcal{E}}(0) = \mathrm{diag}\{\mathcal{E}(0), \quad \mathcal{E}(0), \ldots, \mathcal{E}(0)$$

solves Eq. (2.3.30) in this ring. Further, the equalities

$$(Q_1(z)z - Q_2(z)\bar{z})\Gamma_j(z) = \Gamma_{j-1}(z)$$

(like in the case of the Wronsky matrices) imply that the matrix v has the form

$$v = \begin{pmatrix} v_1 & I & 0 & \ldots & 0 \\ v_2 & 0 & I & & \\ \vdots & \vdots & & & I \\ v_N & 0 & & & 0 \end{pmatrix},$$

with a matrix v_j being found from the system of linear equations ($v_j \in \mathrm{Mat}_r(C^\infty(R^2)$)

$$\sum_{j=1}^{N} \Gamma_j(z_i)v_j = \Gamma_0(z_i) \qquad (1 \leqslant i \leqslant N),$$

Hence,

$$PN^{-1}v = PN^1vP = v_N, \qquad Pv^{-1}N = Pv^{-1}NP = v_N^{-1},$$

and since the matrix $\widetilde{\mathcal{E}}(0)$ is commutative with the matrices P and N,

$$PN^{-1}\widetilde{\mathcal{E}}(0)v = PN^{-1}\widetilde{\mathcal{E}}(0)vP = \mathcal{E}(0)v_N.$$

This enables us to apply a projection operation and to obtain the solution $\mathcal{E}(0)v_N$ of Eq. (2.3.30), belonging to the ring $\mathrm{Mat}_r(C^\infty(R^2))$. According to Corollary 2.2.4, the matrix $S = \rho^{-1}\mathcal{E}(0)v_N$ with $\rho = \prod_{k=1}^N |z_k|$ is unitary, belongs to the ring $K_0 = \mathrm{Mat}_r(C^\infty(R^2))$, and obviously satisfies Eq. (2.3.30) as well.

§4 Singular Solutions of Nonlinear Equations

The solutions of nonlinear equations obtained in the previous section are infinitely differentiable for all real values of variables $x_1, x_2, \ldots$. To meet this condition, one must choose matrices Γ so that their determinants should not equal zero for all real values of these variables. If the determinant is not identically equal to zero, then the corresponding solutions would be infinitely differentiable only in the domain where $\operatorname{Det}\Gamma \neq 0$ and turn to infinity on the boundary of the domain. In this section, we consider some examples of such solutions.

Typical singular solutions of the KdV equation are obtained in the case when the matrix A has a nondiagonal Jordan form in Eq. (2.3.1).

Let

$$A = \{I_{r_1}(\mu_1), I_{r_2}(\mu_2), \ldots, I_{r_2}(\mu_p)\},$$

where

$$I_r(\mu) = \begin{pmatrix} \mu & 1 & 0 & \ldots & 0 \\ 0 & \mu & 1 & & \vdots \\ & & & & 1 \\ 0 & 0 & \ldots & & \mu \end{pmatrix}$$

is a Jordan box of the order r. To make the Wronsky matrix

$$W = W\{\partial; f(r_1, 1), \ldots, f(r_1, r_1), \ldots, f(r_p, 1), \ldots, f(r_p, r_p)\}$$

satisfy Eq. (2.3.1), it is necessary and sufficient to make the vectors

$$f(r_k) = (f(r_k, 1), f(r_k, 2), \ldots, f(r_k, r_{k-1}), f(r_k, r_k))$$

satisfy the equation

$$\frac{\partial}{\partial t}\vec{f}(r_k) + \frac{\partial^3}{\partial x^3}\vec{f}(r_k) = 0, \qquad \frac{\partial^2}{\partial x^2}\vec{f}(r_k) = I_{r_k}(\mu_k)\vec{f}(r_k).$$

The general solution of the equations

$$\frac{\partial}{\partial t}\vec{f}(r) + \frac{\partial^3}{\partial x^3}\vec{f}(r) = 0, \qquad \frac{\partial^2}{\partial x^2}\vec{f}(r) = I_r(\mu)\vec{f}(r)$$

is found by the standard method, with the components of the vector $\vec{f}(r)$ having the form

$$\begin{aligned}
f(r,1,\mu) &= u_1(\mu) + \frac{1}{1!}\frac{\partial}{\partial\mu}u_2(\mu) + \cdots + \frac{1}{(r-1)!}\frac{\partial^{r-1}}{\partial\mu^{r-1}}u_r(\mu), \\
f(r,2,\mu) &= u_2(\mu) + \frac{1}{1!}\frac{\partial}{\partial\mu}u_3(\mu) + \cdots + \frac{1}{(r-2)!}\frac{\partial^{r-2}}{\partial\mu^{r-2}}u_r(\mu), \\
&\vdots \\
f(r,r-1,\mu) &= u_{r-1}(\mu) + \frac{1}{1!}\frac{\partial}{\partial\mu}u_r(\mu), \\
f(r,r,\mu) &= u_r(\mu),
\end{aligned}$$

with

$$u_k(\mu) = a_k \operatorname{ch}\sqrt{\mu}(x-\mu t) + b_k\mu^{-\frac{1}{2}}\operatorname{sh}\sqrt{\mu}(x-\mu t), \qquad a_k, b_k$$

being arbitrary constants.

Thus, if

$$W = W\{\partial; f(r_1,1,\mu_1), f(r_1,2,\mu_1), \ldots, f(r_p,1,\mu_p), \ldots, f(r_p,r_p,\mu_p)\},$$

then the function

$$v(x,t) = -\frac{\partial^2}{\partial x^2} \ln \operatorname{Det} W$$

solves the KdV equation

$$4v_t + v_{xxx} - 12vv_x = 0.$$

The solutions are real if $\operatorname{Im} a_k = \operatorname{Im} b_k = \operatorname{Im} \mu_k = 0$. Rational solutions are obtained in the case when $\mu_1, \mu_2 = \cdots = \mu_p = 0$. With $\mu_k > 0$, one obtains exponentially decreasing solutions; with $\mu_k - \kappa_k^2 (\kappa_k^2 > 0)$, one obtains slowly decreasing, oscillating solutions. For instance, if the matrix A consists of a single Jordan block of the order r, i.e.,

$$A = \begin{pmatrix} \mu & 1 & 0 & 0 \\ 0 & \mu & 1 & \vdots \\ \vdots & \vdots & & 1 \\ 0 & 0 & & \mu \end{pmatrix} \qquad (\mu = -\kappa^2, \kappa^2 > 0),$$

then at $r = 1$, one obtains a singular soliton

$$v_1(x,t) = \frac{\kappa^2}{\cos^2 \kappa(x + \kappa^2 t + c)}.$$

At all other odd values $r = 2n+1$, the solutions obtained have the same asymptotic behavior as $v_1(x,t)$:

$$v_{2n+1}(x,t) = \frac{\kappa^2}{\cos^2 \kappa(x + \kappa^2 t + c)}(1 + 0(x^{-1})) \qquad (x \to \infty).$$

For even values of r, one obtains solutions with another asymptotics:

$$v_{2n}(x,t) = n(n-1)\frac{2\kappa \sin 2\kappa(x + \kappa^2 t + c)}{x + 3\kappa^2 t + c} + 0(x^{-2}).$$

All the above solutions are singular—they have second-order poles on the real axis x. This class of solutions plays an important role in problems of asymptotic behavior of nonsingular slowly decreasing solutions of the *KdV* equation when $x \to \infty$. Davydov, Novikov, and Henkin singled out classes of nonsingular decreasing solutions where leading terms of asymptotic expansions (when $x \to \infty$) are either rational or osciallating solutions of the form described above. Similar facts are true also for other nonlinear equations.

CHAPTER 3

REALIZATION OF THE GENERAL SCHEME IN OPERATOR ALGEBRAS

An algebra (over the field of complex numbers C) is a ring K in which the operation of multiplication by complex numbers is defined; multiplication satisfies the following conditions:

1. $1 \cdot x = x$, $\lambda(x+y) = \lambda x + \lambda y$, $(\lambda + \mu)x = \lambda x + \mu x$, $(\lambda\mu)x = \lambda(\mu x)$
2. $\lambda(xy) = x(\lambda y)$ $\quad (x, y \in K, \quad \lambda, \mu \in C)$.

A typical example of an algebra is provided by the set $B(H)$ of all continuous linear operators in the Hilbert space H with usual arithmetical operations for the operators. The set of all possible mappings of an arbitrary set M into an algebra $B(H)$ also generates an algebra, which we shall denote by $A(M, B(H))$. Elements of the latter are operator functions $\hat{a}(m)(m \in M, \hat{a}(m) \in B(H))$ over which the arithmetical operations are performed according to the usual rules:

$$(\hat{a} + \hat{b})(m) = \hat{a}(m) + \hat{b}(m), \qquad \hat{a}\hat{b}(m) = \hat{a}(m)\hat{b}(m), \qquad (\lambda\hat{a})(m) = \lambda\hat{a}(m).$$

In the capacity of M, we take the direct product $Z \times R^m$ of the set Z of all integers by the m-dimensional Euclidian space R^m. We shall denote points of the set $Z \times R^m$ by $(n; x)$ (or, if necessary, by $(n; x_1, x_2, \ldots, x_m)$); and we shall denote the operator functions mapping $Z \times R^m$ into $B(H)$ by $\hat{a} = \hat{a}(n; x) = \hat{a}(n; x_1, x_2, \ldots, x_m)$. The set of all infinitely differentiable mappings of $Z \times R^m$ into $B(H)$ generates a subalgebra in the algebra $A(Z \times R^m, B(H))$ which we shall denote by $C^\infty(B(H))$. In this algebra, we shall define the operations

$$\partial_i \hat{a}(n; x) = \frac{\partial \hat{a}(n; x)}{\partial x_i}, \qquad \alpha(\hat{a}(n; x)) = \hat{a}(n+1; x),$$
$$\partial_\alpha \hat{a}(n; x) = \hat{a}(n+1; x) - \hat{a}(n; x),$$

which obviously are differentiations $\partial_i (i = 1, 2, \ldots, m)$, an automorphism α and a g.d. $\partial_\alpha = d - I$.

All the results established in Chapter I for an arbitrary associative ring are naturally valid for the algebra $C^\infty(B(H))$ which now will play the role of the ring K.

Operators $\hat{b} \in B(H)$ can be regarded as elements of the algebra $C^\infty(B(H))$ by identifying them with the corresponding operator functions $\hat{a}(n;x) \equiv \hat{b}$. Thus, the algebra $B(H)$ is identified with the subalgebra of the algebra $C^\infty(B(H))$ consisting of the operator functions which are annulled by the differentiations $\partial_i (i = 1, 2, \ldots, m)$ introduced above and the g.d.'s ∂_α. In what follows, we stipulate that the operator functions $\hat{a}(n;x) \in C^\infty(B(H))$ just be called *operators*, while those belonging to the subalgebra $B(H)$ be called *constant operators.*

Let us clarify what a projection operation is in the algebra $C^\infty(B(H))$. Every constant idempotent ($\widehat{P}^2 = \widehat{P}$) operator $\widehat{P} \in B(H)$ is known to be a projector (not necessarily orthogonal) onto a certain subspace H_0 of the space H. So the operators $\widehat{P}\hat{a}(n;x)\widehat{P}$ of which the subalgebra $\widehat{P}C^\infty(B(H))\widehat{P}$ consists belong to the algebra $C^\infty(B(H_0))$, while any operator

$$\hat{a}_0(n;x) \in C^\infty(B(H_o))$$

can be represented in such a form, since

$$\hat{b}(n;x) = \hat{a}_0(n;x)\widehat{P} + I - \widehat{P} \in C^\infty(B(H))$$

and

$$\widehat{P}\hat{b}(n;x)\widehat{P} = \hat{a}_0(n;x).$$

Therefore, the projection operations map the algebra $C^\infty(B(H))$ onto $C^\infty(B(H_0))$:

$$\widehat{P}C^\infty(B(H))\widehat{P} = C^\infty(B(H_0)), \qquad H_0 = \widehat{P}(H).$$

If the subset H_0 is one-dimensional, two-dimensional, and so on, then the algebra $C^\infty(B(H_0))$ is identified in the natural way with

$$C^\infty(Z \times R^m), \qquad \mathrm{Mat}_2(C^\infty(Z \times R^m)),$$

and so on.

In accordance with the general scheme for finding a solution of the considered nonlinear equations in the algebra $C^\infty(B(H_0))$ (in particular, in $C^\infty(Z \times R^m), \mathrm{Mat}_r(C^\infty(Z \times R^m))$ etc.), one has:

1. to extend the initial space H_0 up to the space H and choose a constant operator $\widehat{P} = \widehat{P}^2 \in B(H)$ that projects H onto H_0;
2. to solve the linear equations in the left part of Table II in the algebra $C^\infty(B(H))$;
3. to retain only those solutions that satisfy an additional equation (1.3.3), (1.3.4), or (1.3.5);
4. to select the solutions which are invertible in the algebra $C^\infty(B(H))$ from those retained in the previous step;

5. to find logarithmic derivatives of the selected solutions and perform the projection operation onto the subalgebra $\widehat{P}C^\infty(B(H))\widehat{P} = C^\infty(B(H_0))$.

§1 Extension of Algebra $C^\infty(B(H_0))$

Let $C^\infty(B(H_0))$ be the algebra in which we seek solutions of the nonlinear equations under consideration. We noted more than once that in the most interesting cases the space H_0 is of finite dimensionality, with the result that $C^\infty(B(H_0))$ coincides with a matrix algebra $\mathrm{Mat}_r(C^\infty(Z \times R^m))$. We consider the general case when H_0 is an arbitrary separable Hilbert space with the scalar product $(\cdot,\cdot)_{H_0}$ and the norm $\|\cdot\|_{H_0}$. We shall denote vectors of this space by $\vec{f}, \vec{g}, \vec{h}$, and so on.

We are going to recall some basic concepts of the theory of integrating vector and operator functions. Details can be found in [19].

A measurable space $(\Omega, \mathfrak{A}, \partial\mu)$ with the finite measure is an aggregate consisting of an arbitrary set Ω, of a σ-algebra $\mathfrak{A}$ of subsets Ω, and of a σ-additive measure $\mu(A)(A \in \mathfrak{A})$ defined on the algebra and satisfying the inequality $O < \mu = \mu(\Omega) < \infty$.

By

$$L^1_\mu(\Omega), \qquad L^2_\mu(\Omega), \qquad L^\infty_\mu(\Omega)$$

we shall denote the Banach spaces of measurable numerical functions $f(w)$ ($\omega \in \Omega$) with the norms defined in the following way:

$$\|f\|_1 = \mu^{-1}\int_\Omega |f(w)|d\mu; \qquad \|f\|_2^2 = \mu^{-1}\int_\Omega |f(w)|^2 d\mu;$$

$$\|f\|_\infty = \inf_x\{x \in R^1 : |f(\omega)| \leqslant x, \mu \quad \text{almost everywhere} \quad \omega \in \Omega\}.$$

The space $L^2_\mu(\Omega)$ is a Hilbert space with the scalar product

$$(f,g) = \mu^{-1}\int_\Omega f(\omega)\overline{g(\omega)}d\mu.$$

Since the measure $d\mu$ is finite, all constant functions belong to the set $L^\infty_\mu(\Omega)$, and $L^1_\mu(\Omega) \supset L^2_\mu(\Omega) \supset L^\infty_\mu(\Omega)$.

The vector function $\vec{f}(\omega)(\omega \in \Omega)$ with values from H_0 is called measurable if numerical functions $(\vec{h}, \vec{f}(\omega))_{H_0}$ are measurable for all $\vec{h} \in H_0$. Sets of measurable vector functions meeting the condition $\|\vec{f}(\omega)\|_{H_0} \in L^P_\mu(\Omega)$ are denoted by $L^P_\mu(\Omega, H_0)(p = 1, 2, \infty)$.

The set $L^2_\mu(\Omega, H_0)$ is a Hilbert space with usual operations of addition, multiplication by a number, and the scalar product

$$(\vec{f}(\omega), \vec{g}(\omega))_H = \mu^{-1}\int_\Omega (\vec{f}(\omega), \vec{g}(\omega))_{H_0} d\mu.$$

This set is also called a direct integral of sets H_0 by the normed measure $\mu^{-1}d\mu$. We shall denote this space by H:

$$H = L^2_\mu(\Omega, H_0) = \int_\Omega H_0 \mu^{-1} d\mu.$$

All constant vector functions obviously belong to the space (since the measure is finite) and form a subspace in it which is naturally identified with the subspace H_0.

Therefore, the space H is an extension of the space H_0, while the algebra $C^\infty(B(H))$ is an extension of the algebra $C^\infty(B(H_0))$.

Having chosen an arbitrary orthonormal basis $\{\vec{e}_k\}$ in the space H_0, we can express the function $\vec{f}(\omega) \in H$ in the form

$$\vec{f}(\omega) = \sum f_k(\omega)\vec{e}_k,$$

where

$$f_k(\omega) = (\vec{f}(\omega), \vec{e}_k)_{H_0} \in L^2_\mu(\Omega)$$

and

$$\sum \|f_k(\omega)\|_2^2 < \infty.$$

The operator function $\hat{a}(\omega) \in A(\Omega, B(H_0))$ is called measurable if for all $\vec{h} \in H_0$ vector functions $\hat{a}(\omega)\vec{h}$ are measurable. Sets of measurable operator functions satisfying the condition $\|\hat{a}(\omega)\|_{H_0} \in L^p_\mu(\Omega)$ are denoted by $L^p_\mu(\Omega, B(H_0))(p = 1, 2, \infty)$. Since the measure is finite,

$$L^1_\mu(\Omega, B(H_0)) \supset L^2_\mu(\Omega, B(H_0)) \supset L^\infty_\mu(\Omega, B(H_0)) \supset B(H_0).$$

If $\vec{f}(\omega) \in L^1_\mu(\Omega, H_0)$, then for all $\vec{h} \in H_0, (\vec{h}, \vec{f}(\omega))_{H_0} \in L^1_\mu(\Omega)$, and the integral

$$\int_\Omega (\vec{h}, \vec{f}(\omega))_{H_0} d\mu = l(\vec{h})$$

is the bounded linear functional in the space H_0. Hence, there exists such a vector $\vec{g} \in H_0$ that $l(\vec{h}) = (\vec{h}, \vec{g})$. The vector is called an integral of $\vec{f}(\omega)$ and is denoted by $\int_\Omega \vec{f}(\omega)d\mu$. Thus the equation

$$\int_\Omega (\vec{h}, \vec{f}(\omega))_{H_0} d\mu = (\vec{h}, \int_\Omega \vec{f}(\omega)d\mu)_{H_0}$$

is a definition of the integral of $\vec{f}(\omega) \in L^1_\mu(\Omega, H_0)$. The equality implies, in particular, that

$$\| \int_\Omega \vec{f}(\omega)d\mu\|_{H_0} \leqslant \int_\Omega \|\vec{f}(\omega)\|_{H_0} d\mu. \tag{3.1.1}$$

If $\hat{a}(\omega) \in L^1_\mu(\Omega, B(H_0))$, then, for all $\vec{h} \in H_0$, the vector function $\hat{a}(\omega)\vec{h}$ belongs to the space $L^1_\mu(\Omega, H_0)$, and the integral $\int_\Omega \hat{a}(\omega)\vec{h}d\mu$ is a bounded linear operator in the space H_0. This operator is called the integral of $\hat{a}(\omega)$ and is denoted by $\int_\Omega \hat{a}(\omega)d\mu$. So the equality

$$\int\limits_\Omega \hat{a}(\omega)\vec{h}d\mu = (\int\limits_\Omega \hat{a}(\omega)d\mu)\vec{h}$$

is a definition of the integral of $\hat{a}(\omega) \in L^1_\mu(\Omega, B(H_0))$. In particular, it implies that

$$\|\int\limits_\Omega \hat{a}(\omega)d\mu\|_{H_0} \leqslant \int\limits_\Omega \|\hat{a}(\omega)\|_{H_0}d\mu.$$

Let us single out some operator classes that belong to the algebra $B(H)$. Note, above all, that the operators $\hat{a} \in B(L^2_\mu(\Omega))$ and $\hat{c} \in B(H_0)$ may be regarded as operators from $B(H)$ whose action on vector functions $\vec{f}(\omega) = \sum f_k(\omega)\vec{e}_k \in H$ is given by the equalities

$$\hat{a}(\vec{f}(\omega)) = \sum \hat{a}(f_k(\omega))\vec{e}_k, \qquad \hat{c}(\vec{f}(\omega)) = \sum f_k(\omega)\hat{c}(\vec{e}_k).$$

It is clear that the operators $\hat{a}, \hat{c}$ are commutative and

$$\|\hat{a}\|_H = \|\hat{a}\|_{L^2_\mu(\Omega)}, \qquad \|\hat{c}\|_H = \|\hat{c}\|_{H_0}.$$

Thus, the algebras $B(L^2_\mu(\Omega)), B(H_0)$ are identified with the corresponding subalgebras of the algebra $B(H)$, which in the sequel will be denoted by the same symbols.

The operator functions $\hat{a}(\omega) \in L^\infty_\mu(\Omega, B(H_0))$ also generate operators from $B(H)$ whose action is described by the equality

$$\hat{a}(\omega)(\vec{f}(\omega)) = \sum f_k(\omega)\hat{a}(\omega)(\vec{e}_k),$$

with

$$\|\hat{a}(\omega)\|_H = \|\|\hat{a}(\omega)\|_{H_0}\|_\infty$$

Such operators are called *decomposable* and form in $B(H)$ a subalgebra which we shall denote by $L^\infty_\mu(\Omega, B(H_0))$. Decomposable operators of the form $\hat{a}(\omega) = a(\omega)I$, where $a(\omega) \in L^\infty_\mu(\Omega)$, are called operators of multiplication by the function $a(\omega)$ and are denoted by $a(\omega)$. In $L^\infty_\mu(\Omega, B(H_0))$, they form a subalgebra denoted by $L^\infty_\mu(\Omega)$. The subalgebra of decomposable operators consists of all the operators of the algebra $B(H)$ which commute with all the operators of multiplication by a function $a(\omega) \in L^\infty_\mu(\Omega)$.

Assume now that an operator function

$$\hat{q}(\omega) \in L^2_\mu(\Omega, B(H_0)) \subset L^1_\mu(\Omega, B(H_0)),$$

and that the operator

$$\hat{q} = \int\limits_{\Omega} \hat{q}(\omega) d\mu \in B(H_0)$$

is invertible. Then

$$\|\hat{q}^{-1}\hat{q}(\omega)(\vec{f}(\omega))\|_{H_0} \leqslant \|\hat{q}^{-1}\|_{H_0} \|\hat{q}(\omega)\|_{H_0} \|\vec{f}(\omega)\|_{H_0}$$

and, if $\vec{f}(\omega) \in L^2_\mu(\Omega, H_0)$, then the vector function $\hat{q}^{-1}\hat{q}(\omega)(\vec{f}(\omega))$ belongs to the space $L^1_\mu(\Omega, H_0)$, since

$$\|\hat{q}(\omega)\|_{H_0} \in L^2_\mu(\Omega), \qquad \|\vec{f}(\omega)\|_{H_0} \in L^2_\mu(\Omega).$$

Hence, the integral exists,

$$\int\limits_{\Omega} \hat{q}^{-1}\hat{q}(\omega)(\vec{f}(\omega)) d\mu \in H_0$$

and, according to Eq. (3.1.1) and the Cauchy–Schwarz–Buniakowski inequality

$$\begin{aligned}
\|\int\limits_{\Omega} \hat{q}^{-1}\hat{q}(\omega)(\vec{f}(\omega)) d\mu\|_{H_0} &\leqslant \int\limits_{\Omega} \|\hat{q}^{-1}\hat{q}(\omega)(\vec{f}(\omega))\|_{H_0} d\mu \\
&\leqslant \|\hat{q}^{-1}\|_{H_0} \int\limits_{\Omega} \|\hat{q}(\omega)\|_{H_0} \|\vec{f}(\omega)\|_{H_0} d\mu \\
&\leqslant \|\hat{q}^{-1}\|_{H_0} (\mu \int\limits_{\Omega} \|\hat{q}(\omega)\|_{H_0} d\mu)^{\frac{1}{2}} \|\vec{f}(\omega)\|_{H_0}.
\end{aligned}$$

$$\widehat{P}(\vec{f}(\omega)) = \hat{q}^{-1} \int\limits_{\Omega} \hat{q}(\omega)(\vec{f}(\omega)) d\mu,$$

is bounded in the space H, its domain of values lies in H_0, and, if $\vec{f}(\omega) \equiv \vec{f} \in H_0$, then

$$\widehat{P}(\vec{f}) = \hat{q}^{-1} \int\limits_{\Omega} \hat{q}(\omega)(\vec{f}) d\mu = \hat{q}^{-1} \int\limits_{\Omega} \hat{q}(\omega) d\mu(\vec{f}) = \hat{q}^{-1}\hat{q}(\vec{f}) = \vec{f}.$$

Thus, $\widehat{P}^2 = \widehat{P}$, $\widehat{P}(H) = H_0$, that is, $\widehat{P}$ is a projector on the subspace H_0.

Setting $\hat{p}(\omega) = \hat{q}^{-1}\hat{q}(\omega)$, we find that

$$\widehat{P}(\vec{f}(\omega)) = \int\limits_{\Omega} \hat{p}(\omega)(\vec{f}(\omega)) d\mu, \tag{3.1.2}$$

where $\hat{p}(\omega)$ is an arbitrary operator function satisfying the condition

$$\hat{p}(\omega) \in L^2_\mu(\Omega, B(H_0)), \qquad \int\limits_{\Omega} \hat{p}(\omega) d\mu = I.$$

§2 Solving Linear Equations in the Algebra $C^\infty(B(H))$

In order to solve the linear equations of the left part of Table II in the algebra $C^\infty(B(H))$, the method of separation of variables can be employed. One must bear in mind, however, that, although the equations at hand are linear and homogeneous, the coefficients in them are operators. That is why the superposition principle acquires such a form: solutions can be added (integrated with respect to the parameter) and multiplied on the right (resp. on the left) by constant operators that are commutative with all the right (resp. left) coefficients of the equations considered. In other words, the set of solutions of these equations is the right module over the algebra of constant operators commutative with the right coefficients, and is the left module over the algebra of constant operators commutative with the left coefficients.

Since, in what follows, dimensionality of the space R^m will not exceed three, and the variables x_1, x_2, x_3 will have different physical meaning, we shall henceforth employ such standard notation:

$$x_1 = x, \qquad x_2 = y, \qquad x_3 = t; \qquad \partial_x = \frac{\partial}{\partial x}, \qquad \partial_y = \frac{\partial}{\partial y}, \qquad \partial_t = \frac{\partial}{\partial t}.$$

In the same manner, operators $\widehat{\Gamma} \in C^\infty(B(H))$ will be denoted by $\widehat{\Gamma}(n; x, y, t)$, omitting those variables from n, x, y, t which are inessential for the problem at hand. Detailed solutions will be written out only for typical equations. In all other cases, after referring to the appropriate number in Table II, we shall give only the final expressions for those operators $\widehat{\Gamma}$ and nonlinear equations which are solved by their logarithmic derivatives. In the sequel, we shall ascribe to nonlinear equations and formulas for $\widehat{\Gamma}$ the number corresponding to the one in Table II.

We shall start with equations containing two differentiations.

KdV equation:

By setting

$$\partial = \partial_x, \qquad \partial_0 = \partial_t + \lambda\partial_x \qquad (\lambda \in R^1)$$

in Eq. II(1′), we find that $\hat{\gamma} = \widehat{\Gamma}^1\partial_x\Gamma$ satisfies the KdV equation

$$\partial_x(4\partial_t\hat{\gamma} + 4\lambda\partial_x\gamma + \partial_x^3\hat{\gamma} + 6\partial_x\hat{\gamma}\partial_x\hat{\gamma}) = 0, \tag{3.2.II(1′)}$$

if $\widehat{\Gamma}$ is a compatible solution of the system

$$(\partial_t + \lambda\partial_x + \partial_x^3)\widehat{\Gamma} = \widehat{C}\widehat{\Gamma}, \qquad \partial_x^2\widehat{\Gamma} = \widehat{A}\widehat{\Gamma}.$$

Since differentiations ∂_t and ∂_x are commutative, this system has an invertible compatible solution only if the operators $\widehat{C}$ and $\widehat{A}$ are commutative. So if $\widehat{\Gamma}$ is a compatible solution of this system, then $\widehat{\Gamma}_1 = e^{-\widehat{C}t}\widehat{\Gamma}$ satisfies the equations

$$(\partial_t + \lambda\partial_x + \partial_x^3)\widehat{\Gamma}_1 = 0, \qquad \partial_x^2\widehat{\Gamma}_1 = \widehat{A}\widehat{\Gamma}_1,$$

with

$$\hat{\gamma}_1 = \hat{\Gamma}_1^{-1}\partial_x\hat{\Gamma}_1 = \hat{\Gamma}^{-1}e^{\hat{C}t}e^{-\hat{C}t}\partial_x\hat{\Gamma} = \hat{\Gamma}^{-1}\partial_x\hat{\Gamma} = \hat{\gamma}.$$

This permits us, without loss of generality, to find the compatible solution of the system

$$(\partial_t + \lambda\partial_x + \partial_x^3)\hat{\Gamma} = 0, \qquad \partial_x^2\hat{\Gamma} = \hat{A}\hat{\Gamma}.$$

With $\hat{A} = \hat{a}^2$, the general solution of the latter equation is

$$\hat{\Gamma} = e^{\hat{a}x}\hat{C}_1(t) + e^{-\hat{a}x}\hat{C}_2(t),$$

where $\hat{C}_1(t), \hat{C}_2(t) \in C^\infty(B(H))$ do not depend on x. Substituting this expression into the former equation, we find

$$\hat{C}_1(t) = \exp(-\lambda\hat{a} - \hat{a}^3)t\hat{C}_1, \qquad \hat{C}_2(t) = \exp(\lambda\hat{a} + \hat{a}^3)t\hat{C}_2,$$

whence

$$\hat{\Gamma} = e^{\hat{\theta}}\hat{C}_1 + e^{-\hat{\theta}}\hat{C}_2, \qquad \theta = \hat{a}x - (\lambda\hat{a} + \hat{a}^3)t, \tag{3.2.II(1')}$$

where $\hat{a}, \hat{c}_1, \hat{c}_2$ are arbitrary constant operators from $B(H)$.

Equations of a nonlinear string:
Setting

$$\partial = \partial_x, \qquad \partial_1 = \varepsilon\partial_t, \qquad \lambda = -\frac{3}{4}\delta^2 \qquad (\varepsilon, \delta \text{ are numbers})$$

in Eqs. II(1″), we find that $\hat{\gamma} = \hat{\Gamma}^{-1}\partial_x\hat{\Gamma}$ satisfies the equation

$$-3\delta^2\partial_x^2\hat{\gamma} + \partial_x^4\hat{\gamma} + 6\partial_x(\partial_x\hat{\gamma}\partial_x\hat{\gamma}) + 3\varepsilon^2\partial_t^2\hat{\gamma} + 6\varepsilon[\partial_t\hat{\gamma}, \partial_x\hat{\gamma}] = 0 \tag{3.2.II(1″)}$$

provided that $\hat{\Gamma}$ is a compatible solution of the system

$$(4\partial_x^3 - 3\delta^2\partial_x)\hat{\Gamma} = 4\hat{C}\hat{\Gamma}, \qquad (\varepsilon\partial_t + \partial_x^2)\hat{\Gamma} = \hat{A}\hat{\Gamma}.$$

As in the previous example, we may assume without loss of generality that $\hat{A} = 0$. Using the elementary identity $\cos 3z = 4\cos^3 z - 3\cos z$, we see that the operators $\hat{\Gamma} = \sum_{j=0}^2 \exp(\hat{a}_j x)\hat{C}_j$ will satisfy the former equation of the system if

$$\partial_x\hat{C}_j = 0, \qquad \hat{a}_j = \delta\cos\left(\hat{a} + \frac{2\pi}{3}jI\right), \qquad \hat{a} \in B(H)$$

and $4\hat{C} = \delta^3\cos 3\hat{a}$. Substituting these operators $\hat{\Gamma}$ into the latter equation ($\hat{A} = 0$), we find the dependence of the coefficients $\hat{C}_j$ on t and the final expression for $\hat{\Gamma}$:

$$\hat{\Gamma} = \sum_{j=0}^{2} e^{\hat{\theta}_j}\hat{C}_j, \qquad \hat{\theta}_j = \hat{a}_j x - \varepsilon^{-1}\hat{a}_j^2 t, \tag{3.2.II(1″)}$$

where $\hat{a}_j = \delta\cos(\hat{a} + \frac{2\pi}{3}jI)$ and $\hat{a}, \widehat{C}_0, \widehat{C}_1, \widehat{C}_2$ are arbitrary operators in $B(H)$.

Modified KdV equation (II(2)):
If

$$\widehat{\Gamma} = e^{\hat{\theta}}\widehat{C}\frac{1}{2}(I+\widehat{B}) + \hat{e}^{-\hat{\theta}}\widehat{C}\frac{1}{2}(I-\widehat{B}), \qquad \hat{\theta} = \hat{a}x - \hat{a}^3 t, \tag{3.2.II(2)}$$

where $\hat{a}, \widehat{C}, \widehat{B} \in B(H), \widehat{B}^2 = I$, then $\hat{\gamma} = \widehat{\Gamma}^{-1}\partial_x\widehat{\Gamma}$ satisfies the equation

$$\hat{u} = [\hat{\gamma}, \widehat{B}], \qquad 4\partial_t\hat{u} + \partial_x^3\hat{u} + 3(\hat{u}^2\partial_x\hat{u} + \partial_x\hat{u}\hat{u}^2) = 0. \tag{3.2.II(2)}$$

Nonlinear Schrödinger equation and Heisenberg equation (II(3′) *and* II(5)):
If

$$\widehat{\Gamma} = e^{\hat{\theta}}\widehat{C}\frac{1}{2}(I+\widehat{B}) + e^{-\hat{\theta}}\widehat{C}\frac{1}{2}(I-\widehat{B}), \qquad \hat{\theta} = \hat{a}x - \varepsilon^{-1}\hat{a}^2 t, \tag{3.2.II(3′)}$$

where $\hat{a}, \hat{C}, \widehat{B} \in B(H), B^2 = I, \varepsilon$ is a number, then $\hat{\gamma} = \widehat{\Gamma}^{-1}\partial_x\widehat{\Gamma}$ satisfies the equation

$$\hat{u} = [\hat{\gamma}, \widehat{B}], \qquad 2\varepsilon\partial_t\hat{u}\widehat{B} + \partial_x^2\hat{u} + 2\hat{u}^3 = 0 \tag{3.2.II(3′)}$$

and another equation

$$\hat{S} = \hat{\gamma}\widehat{B}\hat{\gamma}^{-1}, \qquad 4\varepsilon\partial_t\hat{S} = [\hat{S}, \partial_x^2\hat{S}], \tag{3.2.II(5)}$$

provided that the operator $\hat{a}$ is invertible.

Sine-Gordon equation (II(7′)):
Setting $\partial = \partial_x, \partial_1 = \partial_y$ in the equation II(7′) and taking for α the inner automorphism $\alpha(\widehat{\Gamma}) = \widehat{B}\widehat{\Gamma}\widehat{B}$, where $\widehat{B} \in B(H)$ and $\widehat{B}^2 = I$, we find that $\hat{\gamma} = \widehat{\Gamma}^{-1}\partial_x\widehat{\Gamma}$ satisfies the equation

$$\partial_y(\hat{\gamma}^{-1}\partial_x\hat{\gamma}) = \hat{\gamma}^{-1}\widehat{B}\hat{\gamma}\widehat{B} - \widehat{B}\hat{\gamma}^{-1}\widehat{B}\hat{\gamma}, \tag{3.2.II(7′)}$$

provided that $\widehat{\Gamma}$ is a compatible solution of the system

$$\partial_y\partial_x\widehat{\Gamma} = \widehat{\Gamma}, \qquad \partial_x\widehat{\Gamma} = \widehat{A}\widehat{B}\widehat{\Gamma}\widehat{B}$$

and the operator $\widehat{A} \in B(H)$ is invertible. The general solution of the latter equation has such a form

$$\widehat{\Gamma} = e^{\hat{a}x}\widehat{C}_1(I+\widehat{B}) + e^{-\hat{a}x}\widehat{C}_2(I-\widehat{B}),$$

where $\hat{a} = \widehat{A}\widehat{B}$. Substituting this expression into the former equation, we find the dependence of the coefficients $\widehat{C}_j$ on y: $\widehat{C}_1(y) = e^{\hat{a}^{-1}y}\widehat{C}_1, \widehat{C}_2(y) = e^{-\hat{a}^{-1}y}\widehat{C}_2$, whence

$$\widehat{\Gamma} = e^{\hat{\theta}}\widehat{C}\frac{I+\widehat{B}}{2} + e^{-\hat{\theta}}\widehat{C}\frac{I-\widehat{B}}{2}, \qquad \hat{\theta} = \hat{a}x - \hat{a}^{-1}y, \tag{3.2.II(7′)}$$

where $\hat{a}$ and $\widehat{C} = 2\{\widehat{C}_1(I+\widehat{B}) + \widehat{C}_2(I-\widehat{B})\}$ are arbitrary operators from $B(H)$.

Toda lattice:
Setting

$$\partial = \partial_t, \alpha(\widehat{\Gamma}(n;t)) = \widehat{\Gamma}(n+1;t), \qquad \partial_\alpha(\widehat{\Gamma}(n;t)) = \widehat{\Gamma}(n+1;t) - \widehat{\Gamma}(n;t)$$

in the equations II(6), we find that

$$\hat{\gamma} = \hat{\gamma}(n;t) = \widehat{\Gamma}^{-1}\partial_t\widehat{\Gamma} = \widehat{\Gamma}(n;t)^{-1}\widehat{\Gamma}(n+1;t)$$

satisfies the equation

$$\partial_t(\hat{\gamma}(n)^{-1}\partial_t\hat{\gamma}(n)) = \hat{\gamma}(n)^{-1}\hat{\gamma}(n+1) - \hat{\gamma}(n-1)^{-1}\hat{\gamma}(n), \tag{3.2.II(6)}$$

provided that $\widehat{\Gamma}$ is a compatible solution of the system

$$\partial_t\widehat{\Gamma}(n) = \widehat{\Gamma}(n+1), \qquad \widehat{\Gamma}(n+2) + \widehat{\Gamma}(n) = \widehat{A}\widehat{\Gamma}(n+1).$$

At $\widehat{A} = \hat{a} + \hat{a}^{-1}$, the general solution of the latter equation has the form

$$\widehat{\Gamma} = \hat{a}^n\widehat{C}_1(t) + \hat{a}^{-n}\widehat{C}_2(t).$$

Substituting this expression into the former equation, we see that the operators

$$\widehat{\Gamma} = \hat{a}^n e^{\hat{a}t}\widehat{C}_1 + \hat{a}^{-n}e^{\hat{a}^{-1}t}\widehat{C}_2 \tag{3.2.II(6)}$$

satisfy both equations of the system, provided that $\hat{a}, \widehat{C}_1, \widehat{C}_2 \in B(H)$.

Longmuir lattice (II(4) *or* II(8)):
If

$$\widehat{\Gamma} = \hat{a}^n e^{\hat{a}^2 t}\widehat{C}_1 + \hat{a}^{-n}e^{\hat{a}^{-2}t}\widehat{C}_2, \tag{3.2.II(4)}$$

where $\hat{a}, \widehat{C}_1, \widehat{C}_2 \in B(H)$, then

$$\hat{\gamma}(n;t) = \widehat{\Gamma}^{-1}\partial_t\widehat{\Gamma} = \widehat{\Gamma}(n;t)^{-1}\widehat{\Gamma}(n+2,t)$$

satisfies the equation

$$\begin{gathered} \hat{u}(n;t) = I + \hat{\gamma}(n+1;t) - \hat{\gamma}(n;t) = \hat{\gamma}(n-1;t)^{-1}\hat{\gamma}(n;t) \\ \partial_t\hat{u}(n) = \hat{u}(n)\hat{u}(n+1) - \hat{u}(n-1)\hat{u}(n). \end{gathered} \tag{3.2.II(4)}$$

N-wave problem:
Setting $\partial_1 = \partial_x, \partial_2 = \partial_y$ in the equation II(9′), we find that the logarithmic derivatives $\hat{\gamma}_1 = \widehat{\Gamma}^{-1}\partial_x\widehat{\Gamma}$, $\hat{\gamma}_2 = \widehat{\Gamma}^{-1}\partial_y\widehat{\Gamma}$ satisfy the equations

$$\begin{gathered} \hat{v}_j = [\hat{\gamma}_j, \widehat{D}_j]\widehat{D}_j^{-1} \qquad (j = 1, 2) \\ [\hat{v}_1, \widehat{D}_2] = [\hat{v}_2, \widehat{D}_1], \qquad \partial_y\hat{v}_1 - \partial_x\hat{v}_2 = [\hat{v}_1, \hat{v}_2], \end{gathered} \tag{3.2.II(9′)}$$

provided that $\widehat{\Gamma}$ is a compatible solution of the simultaneous equations

$$\partial_x\widehat{\Gamma} = \hat{a}\widehat{\Gamma}\widehat{D}_1, \qquad \partial_y\widehat{\Gamma} = \hat{a}\widehat{\Gamma}\widehat{D}_2,$$

where

$$\hat{a} \in B(H), \qquad \partial_x\widehat{D}_2 = \partial_y\widehat{D}_1 = 0 \qquad [\widehat{D}_1, \widehat{D}_2] = 0.$$

Note also that simultaneous equations (3.2.II(9′)) are equivalent to a single equation in $\hat{u} = \hat{\gamma}_1\widehat{D}_1^{-1} = \hat{\gamma}_2\widehat{D}_2^{-1}$:

$$\partial_y[\hat{u}, \widehat{D}_1] - \partial_x[\hat{u}, \widehat{D}_2] = [[\hat{u}, \widehat{D}_1], [\hat{u}, \widehat{D}_2]].$$

In the case when the operators $\widehat{D}_1, \widehat{D}_2$ are constant, the general solution of the system can be found by expansion in powers of x, y. It will assume the form

$$\widehat{\Gamma} = \sum_{m,n=0}^{\infty} \frac{(\hat{a})^{n+m}\widehat{C}(\widehat{D}_1x)^n(\widehat{D}_2y)^m}{n!\,m!},$$

where $\widehat{C}$ is an arbitrary operator in $B(H)$. An immediate check shows that the operators

$$\widehat{\Gamma} = \sum_{m,n=0}^{\infty} \frac{(\hat{a})^{n+m}\widehat{C}(\hat{d}_1(x))^n(\hat{d}_2(y))^m}{n!\,m!} \tag{3.2.II(9′)}$$

also satisfy this system, provided that the operator functions $\hat{d}_1(x), \hat{d}_2(y)$ take on values from the same commutative subalgebra of the algebra $B(H)$, with $\widehat{C}$ being an arbitrary operator from $B(H)$, and the operators

$$\partial_x\hat{d}_1(x) = \widehat{D}_1 = \widehat{D}_1(x), \qquad \partial_y\hat{d}_2(y) = \widehat{D}_2 = \widehat{D}_2(y)$$

being invertible in the algebra $C^\infty(B(H))$.

Equations of the theory of chiral fields:

Setting

$$\partial_1 = \partial_x, \qquad \partial_2 = \partial_y, \qquad \lambda_1 = \lambda_2 = (2i)^{-1}$$

in the equations II(10), we find that the logarithmic derivatives

$$\hat{\gamma}_1 = \widehat{\Gamma}^{-1}\partial_x\widehat{\Gamma}, \qquad \hat{\gamma}_2 = \widehat{\Gamma}^{-1}\partial_y\widehat{\Gamma}$$

satisfy the equations

$$\begin{aligned} \hat{u}_1 &= \hat{\gamma}_1\widehat{D}_1\hat{\gamma}_1^{-1}, & \hat{u}_2 &= \hat{\gamma}_2\widehat{D}_2\widehat{\Gamma}_2^{-1}, \\ \partial_y\hat{u}_1 &= \frac{1}{2i}[\hat{u}_1, \hat{u}_2], & \partial_x\hat{u}_2 &= \frac{1}{2i}[\hat{u}_2, \hat{u}_1], \end{aligned} \tag{3.2.II(10)}$$

provided that $\widehat{\Gamma}$ is a compatible solution of the system

$$\partial_x\widehat{\Gamma} = \widehat{A}_1\widehat{\Gamma}\widehat{D}_1, \qquad \partial_y\widehat{\Gamma} = \widehat{A}_2\widehat{\Gamma}\widehat{D}_2, \tag{3.2.1}$$

where

$$\widehat{A}_1, \widehat{A}_2 \in B(H), \qquad \partial_y \widehat{D}_1 = \partial_x \widehat{D}_2 = 0, \qquad [\widehat{D}_1, \widehat{D}_2] = 0,$$

the operators $\widehat{A}_1, \widehat{A}_2, \widehat{D}_1, \widehat{D}_2$ are invertible in the algebra $C^\infty(B(H))$, and

$$2iI = \widehat{A}_1^{-1} + \widehat{A}_2^{-1}.$$

According to the latter equality, $\widehat{A}_1^{-1} - iI = iI - \widehat{A}_2^{-1} = i\hat{a}$ and, therefore,

$$\widehat{A}_1 = -i(I + \hat{a})^{-1}, \qquad \widehat{A}_2 = -i(I - \hat{a})^{-1}, \qquad [\widehat{A}_1, \widehat{A}_2] = 0.$$

From this follows, like in the previous example, that Eqs. (3.2.1) are solved by the operators

$$\text{(3.2.II(10))} \qquad \widehat{\Gamma} = \sum_{m,n=0}^{\infty} \frac{\widehat{A}_1^n \widehat{A}_2^m \widehat{C}(\hat{d}_1(x))^n (\hat{d}_2(y))^m}{n!\, m!}$$
$$\widehat{A}_1 = -i(I + \hat{a})^{-1}, \qquad \widehat{A}_2 = -i(I - \hat{a})^{-1},$$

where $\hat{a}, \widehat{C}$ are arbitrary operations from $B(H)$, the operator-valued functions $\hat{d}_1(x), \hat{d}_2(x)$ take on their values from the same commutative subalgebra of the algebra $B(H)$, and the operators

$$\partial_x(\hat{d}_1(x)) = \widehat{D}_1(x), \qquad \partial_y(\hat{d}_2(y)) = \widehat{D}_2(y)$$

are invertible in the algebra $C^\infty(B(H))$.

Note that in all the above examples we found the solutions that depended on the finite number of parameters from the algebra $B(H)$, that is, on the finite number of constant operators. Consider now the equations that contain three differentiations. In this case, solutions of the corresponding linear equations also depend on the finite number of parameters, but these parameters will be measurable operator functions belonging to the algebra $A(\Sigma, B(H))$, where $\{\Sigma, \mathfrak{A}, \partial\nu\}$ is an arbitrary measurable space.

KP equation:

Setting $\partial_0 = \partial_t$, $\partial = \partial_x$, $\partial_1 = \varepsilon\partial_y$ (ε is a number) in the equations II(1), we find that the logarithmic derivative $\hat{\gamma} = \widehat{\Gamma}^{-1}\partial_x\widehat{\Gamma}$ satisfies the equation

$$\text{(3.2.II(1))} \qquad \partial_x(4\partial_t\hat{\gamma} + \partial_x^3\hat{\gamma} + 6\partial_x\hat{\gamma}\partial_x\hat{\gamma}) + 3\varepsilon^2\partial_y^2\hat{\gamma} + 6\varepsilon[\partial_y\hat{\gamma}, \partial_x\hat{\gamma}] = 0,$$

provided that $\widehat{\Gamma}$ is a compatible solution of the system

$$(\partial_t + \partial_x)\widehat{\Gamma} = \widehat{C}\widehat{\Gamma}, \qquad (\varepsilon\partial_y + \partial_x^2)\widehat{\Gamma} = \widehat{A}\widehat{\Gamma}.$$

Since the operators $(\partial_t + \partial_x^3)$ and $(\varepsilon\partial_y + \partial_x^2)$ are commutative, this system has an invertible and compatible solution only if the operators C and A are commutative. So, if $\widehat{\Gamma}$ satisfies this system of equations, then the operators $\widehat{\Gamma}_1 = \exp(-\widehat{C}t - \varepsilon^{-1}\widehat{A}y)\widehat{\Gamma}$ satisfy the equations

$$(\partial_t + \partial_x^3)\widehat{\Gamma}_1 = 0, \qquad (\varepsilon\partial_y + \partial_x^2)\widehat{\Gamma}_1 = 0,$$

with $\hat{\gamma} = \widehat{\Gamma}^{-1}\partial_x\widehat{\Gamma} = \widehat{\Gamma}_1^{-1}\partial_x\widehat{\Gamma}_1$. Therefore, without loss of generality, we may seek compatible solutions of the system

$$(\partial_t + \partial_x^3)\widehat{\Gamma} = 0, \qquad (\varepsilon\partial_y + \partial_x^2)\widehat{\Gamma} = 0$$

that does not contain any operator coefficients at all.

It is clear that the operators $\exp(\hat{a}x - \hat{a}^3t - \varepsilon^{-1}\hat{a}^2y)$ with any $\hat{a} \in B(H)$ satisfy this system, and from here, using the superposition principle, we find that the operators

$$\widehat{\Gamma} = \int_\Sigma \widehat{C}_1(\xi)e^{\hat{\theta}(\xi)}\widehat{C}_2(\xi)d\nu(\xi),$$
$$\hat{\theta}(\xi) = \hat{a}(\xi)x - \hat{a}(\xi)^3t - \varepsilon^{-1}\hat{a}(\xi)^2y, \tag{3.2.II(1)}$$

for any

$$\hat{a}(\xi) \in L_\nu^\infty(\Sigma), \qquad \widehat{C}_1(\xi), \widehat{C}_2(\xi) \in L_\nu^2(\Sigma),$$

satisfy the same system of equations.

Davy–Stewardson equations (II(3)):

If

(3.2.II(3))

$$\widehat{\Gamma} = \int_\Sigma \{\widehat{C}_1(\xi)e^{\hat{\theta}_1(\xi)}\widehat{C}(\xi)(I + \widehat{B}) + \widehat{C}_2(\xi)e^{\hat{\theta}_2(\xi)}\widehat{C}(\xi)(I - \widehat{B})\}d\nu(\xi),$$

$$\hat{\theta}_1(\xi) = \hat{a}(\xi)x - \delta\hat{a}^2(\xi)t - \varepsilon\hat{a}(\xi)y, \qquad \hat{\theta}_2(\xi) = -\hat{a}(\xi)x + \delta\hat{a}^2(\xi)t - \varepsilon\hat{a}(\xi)y,$$

where

$$\hat{a}(\xi) \in L_\nu^\infty(\Sigma), \qquad \widehat{C}_1(\xi), \widehat{C}_2(\xi), \widehat{C}(\xi) \in L_\nu^2(\Sigma),$$

ε, δ are numbers, then the logarithmic derivative $\hat{\gamma} = \widehat{\Gamma}^{-1}\partial_x\widehat{\Gamma}$ satisfies the system of equations

$$\hat{u} = [\hat{\gamma}, \widehat{B}], \qquad \hat{v} = \{\hat{\gamma}, B\},$$
$$2\delta^{-1}\partial_t\hat{u}\widehat{B} + (\partial_x^2 + \varepsilon^{-1}\partial_y^2)\hat{u} + 2\hat{u}^3 - 2\varepsilon^{-1}\{\hat{u}, \partial_y\hat{v}\} = 0, \tag{3.2.II(3)}$$
$$\varepsilon^{-1}\partial_y\hat{v} + \partial_x\hat{v}\widehat{B} = \hat{u}^2.$$

Two-dimensional Toda lattice (II(7)):

If

$$\widehat{\Gamma} = \widehat{\Gamma}(n; x, y) = \int_\Sigma \widehat{C}_1(\xi)(\hat{a}(\xi))^n \exp\{\hat{a}(\xi)x + \hat{a}(\xi)^{-1}y\}\widehat{C}_2(\xi)d\nu(\xi) \tag{3.2.II(7)}$$

where

$$\hat{a}(\xi), \hat{a}(\xi)^{-1} \in L_\nu^\infty(\Sigma), \qquad \widehat{C}_1(\xi), \widehat{C}_2(\xi) \in L_\nu^2(\Sigma),$$

then the logarithmic derivative $\hat{\gamma}(n) = \hat{\gamma}(n; x, y) = \widehat{\Gamma}^{-1}\partial_x\widehat{\Gamma}$ satisfies the equation

$$\partial_y(\hat{\gamma}(n)^{-1}\partial_x\hat{\gamma}(n)) = \hat{\gamma}(n)^{-1}\hat{\gamma}(n+1) - \hat{\gamma}(n-1)^{-1}\hat{\gamma}(n). \tag{3.2.II(7)}$$

Two-dimensional N-wave problem (II(9)):

If

$$\widehat{\Gamma} = \int_\Sigma \widehat{C}_1(\xi)e^{\varepsilon\hat{a}(\xi)t}\widehat{\mathcal{E}}(\xi, x, y)d\nu(\xi),$$
$$\widehat{\mathcal{E}}(\xi, x, y) = \sum_{m,n=0}^{\infty} \frac{\hat{a}(\xi))^{m+n}\widehat{C}(\xi)(\hat{d}_1(x))^m(\hat{d}_2(y))^n}{n!\,m!}, \tag{3.2.II(9)}$$

where ε is a number, $\hat{a}(\xi) \in L_\nu^\infty(\Sigma), \hat{C}_1(\xi), \widehat{C}_2(\xi) \in L_\nu^2(\Sigma)$, the operator functions $\hat{d}_1(x), \hat{d}_2(x) \in C^\infty(B(H))$ take their values from the same commutative subalgebra of the algebra $B(H)$, and the operators

$$\partial_x(\hat{d}_1(x)) = \widehat{D}_1(x), \qquad \partial_y(\hat{d}_2(y)) = \widehat{D}_2(y)$$

are invertible in the algebra $C^\infty(B(H))$, then the logarithmic derivative $\hat{\gamma} = \widehat{\Gamma}^{-1}\partial_t\widehat{\Gamma}$ satisfies the equation

$$\begin{aligned}\partial_y[\hat{\gamma}, \widehat{D}_1] - \partial_x[\hat{\gamma}, \widehat{D}_2] - \varepsilon^{-1}\partial([\hat{\gamma}, \widehat{D}_1]\widehat{D}_2 - [\hat{\gamma}, \widehat{D}_2]\widehat{D}_1) \\ = [[\hat{\gamma}, \widehat{D}_1], [\hat{\gamma}, \widehat{D}_2]].\end{aligned} \tag{3.2.II(9)}$$

§3 Additional Equations

For performing the projection operation, one must select the operator parameters contained in the previously found solutions of the linear equations in such a way that these solutions would satisfy the corresponding additional equations (1.3.3), (1.3.4), or (1.3.5). This task is reduced to solving one standard operator algebraic equation.

Consider those equations of Table II for which the additional equation is (1.3.3). In the algebra $C^\infty(B(H))$, it assumes the form:

$$\partial\widehat{\Gamma}(I - \widehat{P}) = \widehat{\Gamma}\widehat{N}(I - \widehat{P}),$$

where $\widehat{P}$ is a projection operator onto the subspace H_0 defined by equality (3.1.2), and $\widehat{N}$ is an arbitrary operator from the algebra $B(H)$.

From formula (3.2.II(1)) that yields solutions of the linear equations generating the KP equation, it follows that this additional equation will be satisfied if

$$(\hat{a}(\xi)\widehat{C}_2(\xi) - \widehat{C}_2(\xi)\widehat{N})(I - \widehat{P}) = 0. \tag{3.3.II(1)}$$

Turning to the solutions $\widehat{\Gamma}$ of other equations from this part of Table II (the solutions were given in the previous section), we find that in order to satisfy the additional equations, one must choose the operator parameters so that they would satisfy the equations:

$$\text{(3.3.II(1))} \quad (\hat{a}\widehat{C}_1 - \widehat{C}_1\widehat{N})(I - \widehat{P}) = 0, \quad (-\hat{a}\widehat{C}_2 - \widehat{C}_2\widehat{N})(I - \widehat{P}) = 0;$$

$$\text{(3.3.II(1''))} \quad (\hat{a}_j\widehat{C}_j - \widehat{C}_j\widehat{N})(I - \widehat{P}) = 0, \quad j = 0, 1, 2;$$

$$\text{(3.3.II(3))} \quad (\hat{a}(\xi)\widehat{C}(\xi) - \widehat{C}(\xi)\widehat{B}\widehat{N})(I - \widehat{P}) = 0;$$

$$\text{(3.3.II(2, 3', 7'))} \quad (\hat{a}\widehat{C} - \widehat{C}\widehat{B}\widehat{N})(I - \widehat{P}) = 0;$$

$$\text{(3.3.II(7))} \quad (\hat{a}(\xi)\widehat{C}_2(\xi) - \widehat{C}_2(\xi)\widehat{N})(I - \widehat{P}) = 0;$$

$$\text{(3.3.II(6))} \quad (\hat{a}\widehat{C}_1 - \widehat{C}_1\widehat{N})(I - \widehat{P}) = 0, \quad (\hat{a}^{-1}\widehat{C}_2 - \widehat{C}_2\widehat{N})(I - \widehat{P}) = 0;$$

$$\text{(3.3.II(4))} \quad (\hat{a}^2\widehat{C}_1 - \widehat{C}_1\widehat{N})(I - \widehat{P}) = 0, \quad (\hat{a}^{-2}\widehat{C}_2 - \widehat{C}_2\widehat{N})(I - \widehat{P}) = 0.$$

(To these equations, we ascribe numbers coinciding with those for the corresponding solutions $\widehat{\Gamma}$ from the previous section).

Further, from formulas (3.2.II(9′)) and (3.2.II(9)) it follows that for the N-wave problem and the two-dimensional generalization of it, the additional conditions are met if

$$[\hat{d}_1(x), \widehat{N}] = [\hat{d}_2(y), \widehat{N}] = 0$$

and

$$\text{(3.3.II(9'))} \quad (\hat{a}\widehat{C} - \widehat{C}\widehat{N})(I - \widehat{P}) = 0,$$

$$\text{(3.3.II(9))} \quad (\varepsilon\hat{a}(\xi)\widehat{C}(\xi) - \widehat{C}(\xi)\widehat{N})(I - \widehat{P}) = 0.$$

Finally, consider the chiral field equations. In this case, the conditional equations are as follows

$$\partial_x(\widehat{\Gamma})\widehat{N}_1(I - \widehat{P}) = \widehat{\Gamma}(I - \widehat{P}), \qquad \partial_y(\widehat{\Gamma})\widehat{N}_2(I - \widehat{P}) = \widehat{\Gamma}(I - \widehat{P}),$$

where $\widehat{N}_1, \widehat{N}_2$ are arbitrary invertible operators in $C^\infty(B(H))$ that commute with the operators $\hat{d}_1, (x), \hat{d}_2(y)$:

$$[\hat{d}_1(x), \widehat{N}_1] = [\hat{d}_1(x), \widehat{N}_2] = [\hat{d}_2(y), \widehat{N}_1] = [\hat{d}_2(y), \widehat{N}_2] = 0.$$

Formula (3.2.II(10)) implies that in order to make the operators $\widehat{\Gamma}$ satisfy the additional equations, one must choose the operator parameters in them so that the equalities

$$(\widehat{A}_1\widehat{C}\widehat{D}_1\widehat{N}_1 - \widehat{C})(I - \widehat{P}) = 0, \qquad (\widehat{A}_2\widehat{C}\widehat{D}_2\widehat{N}_2 - \widehat{C})(I - \widehat{P}) = 0,$$

where

$$\widehat{D}_1 = \partial_x(\hat{d}_1(x), \qquad \widehat{D}_2 = \partial_y(\hat{d}_2(y)),$$

should be valid. Since the operators $\widehat{A}_1, \widehat{A}_2$ are invertible and

$$\widehat{A}_1^{-1} = i(I + \hat{a}), \qquad \widehat{A}_2^{-1} = i(I - \hat{a}),$$

these equalities are equivalent to

$$(\hat{a}\widehat{C} - \widehat{C}(-I - i\widehat{D}_1\widehat{N}_1))(I - \widehat{P}) = 0, \qquad (\hat{a}\widehat{C} - \widehat{C}(I + i\widehat{D}_2\widehat{N}_2))(I - \widehat{P}) = 0.$$

Setting

$$-I - i\widehat{D}_1\widehat{N}_1 = I + i\widehat{D}_2\widehat{N}_2 = \widehat{N},$$

we arrive at the equation

$$(\hat{a}\widehat{C} - \widehat{C}\widehat{N})(I - \widehat{P}) = 0, \tag{3.2.II(10)}$$

where $\widehat{N}$ is an arbitrary operator in $C^{\infty}(B(H))$ that satisfies the conditions

$$[\hat{d}_1(x), \widehat{N}] = [\hat{d}_2(y), \widehat{N}] = 0,$$

where the operators $I + \widehat{N}$ and $I - \widehat{N}$ are invertible.

Thus, being given the operators $\hat{a}$ (or $\hat{a}(\xi)$) and $\widehat{N}$ whose choice remains arbitrary, the problem at hand is in all cases reduced to solving in $\widehat{X} \in B(H)$ the following standard equation

$$(\widehat{A}\widehat{X} - \widehat{X}\widehat{N})(I - \widehat{P}) = 0. \tag{3.3.1}$$

Before solving it, we note, first of all, that Eq. (3.3.1) is equivalent to

$$\widehat{A}\widehat{X} - \widehat{X}\widehat{N} = \widehat{R}\widehat{P}, \tag{3.3.2}$$

where $\widehat{R}$ is an arbitrary operator in $B(H)$.

To begin with, we consider the case of one-dimensional space H_0. Then it is identified in a natural manner with the set of all complex numbers, while the space H is identified with the set of scalar functions $f(\omega) \in L^2_{\mu}(\Omega)$. The projection operators onto $H_0 \subset H$ are given by the formula

$$\widehat{P}(f(\omega)) = \int_{\Omega} p(\omega')f(\omega')d\mu(\omega'), \tag{3.3.3}$$

where $p(\omega)$ is an arbitrary function from $L^2_{\mu}(\Omega)$ that satisfies the condition

$$\int_{\Omega} p(\omega)d\mu(\omega) = 1.$$

Every operator $\widehat{R} \in B(H)$ transforms the function $g(\omega) \equiv 1$ into a function $r(\omega) \in L^2_{\mu}(\Omega)$, and hence, it follows from (3.3.3), that the operators assume the form

$$\widehat{R}\widehat{P}(f(\omega)) = \int_{\Omega} r(\omega)p(\omega')f(\omega')d\mu(\omega'),$$

i.e., they are integral operators with degenerated kernels $r(\omega)p(\omega^1)$. In the capacity of $\widehat{A}$ and $\widehat{N}$, we shall take operators of multiplication by a function $A(\omega)$ and $N(\omega)$, belonging to $L^\infty_\mu(\Omega)$, and we shall look for the solution of Eq. (3.3.2) in the form of integral operators with kernels $X_1(\omega,\omega')$. Then we shall arrive at the following equation for the kernel $X_1(\omega,\omega')$:

$$A(\omega)X_1(\omega,\omega') - X_1(\omega,\omega')N(\omega') = r(\omega)p(\omega'),$$

whence

$$X_1(\omega,\omega') = r(\omega)[A(\omega) - N(\omega')]^{-1}p(\omega'),$$

and

$$\widehat{X}_1(f(\omega)) = \int_\Omega r(\omega)[A(\omega) - N(\omega')]^{-1}p(\omega')f(\omega')d\mu(\omega'). \tag{3.3.4}$$

This formula gives, of course, only a formal solution of Eq. (3.3.2), because of the difference $A(\omega) - N(\omega')$ can go to zero for some values of the parameters involved. To avoid this, it is necessary to impose some constraints on the functions $r(\omega), p(\omega'), A(\omega), N(\omega')$ and measure $d\mu(\omega')$; the constraints must guarantee that the right-hand side of Eq. (3.3.4) would be a bounded operator in the space $L^2_\mu(\Omega)$. The exact form of these constraints we shall find later, for each concrete case; for the time being, we assume that the right-hand side of Eq. (3.3.4) correctly determines a bounded operator which is a particular solution of Eq. (3.3.2). We also assume that for each value of ω the set $N^{-1}\{A(\omega)\} = \{\omega' : N(\omega') = A(\omega)\}$ has a zero measure, since this condition in any case, is necessary for the correct definition of the operator in Eq. (3.3.4).

The general solution of Eq. (3.3.2) is equal to the sum of the particular solution obtained and to the general solution of the corresponding homogeneous equation. Proceeding formally, we shall seek the solution of the homogeneous equation also in the form of an integral operator with the kernel $X_0(\omega,\omega')$. For this kernel, we have an equation

$$X_0(\omega,\omega')[A(\omega) - N(\omega')] = 0,$$

which means that for each value of ω the kernel is a generalized function concentrated on the set $N^{-1}\{A(\omega)\}$ of zero measure. Hence, the operator $\widehat{X}_0$ defined by the equality

$$\widehat{X}_0(f(\omega)) = \int_\Omega X_0(\omega,\omega')\chi(\omega,\omega')f(\omega')d\mu(\omega'),$$

where $\chi(\omega,\omega')$ is a characteristic function of the set $N^{-1}\{A(\omega)\}$ and $X_0(\omega,\omega')$ is an arbitrary generalized function, is a formal solution of the considered homogeneous equation.

Thus, if the subset H_0 is one-dimensional and if in the capacity of $\widehat{A}$ we take operators of multiplication by a function $A(\omega), N(\omega) \in L^\infty_\mu(\Omega)$ that satisfy the condition $\mu(N^{-1}\{A(\omega)\}) = 0$, then the formal solutions $\hat{X}$ of Eq. (3.3.1) have the form

(3.3.4′)
$$\hat{X}(f(\omega)) = \int_\Omega X_0(\omega,\omega')\chi(\omega,\omega')f(\omega')d\mu(\omega') + \int_\Omega r(\omega)[A(\omega) - N(\omega')]^{-1}p(\omega')f(\omega')d\mu(\omega').$$

In the general case, when H_0 is an arbitrary separable Hilbert space, formal solutions of Eq. (3.3.1) are found in a similar manner. In the capacity of $\hat{A}, \hat{N}$ we take decomposable operators

$$\hat{A} = \sum_i a_i(\omega)\hat{Q}_i, \qquad \hat{N} = \sum_i n_i(\omega)\hat{Q}_i,$$

that satisfy the following conditions:

1) the operators $\hat{Q}_i \in B(H_0)$ are orthogonal projectors, $\sum_i \hat{Q}_i = I$;
2) the scalar functions $a_i(\omega), n_i(\omega)$ belong to the space $L^\infty_\mu(\Omega)$, and for each value of ω the sets $\Omega_{ij} = \{\omega' : n_j(\omega') = a_i(\omega)\}$ have the zero measure.

Then the formal solution of Eq. (3.3.1) are the "integral" operators $\hat{X}$ of the form

(3.3.5)
$$\hat{X}(\vec{f}(\omega)) = \sum_{i,j}\int_\Omega \hat{Q}_i[\hat{X}_0(\omega,\omega')\chi_{ij}(\omega,\omega') + \frac{\hat{r}(\omega)\hat{p}(\omega')}{a_i(\omega) - n_j(\omega')}]\hat{Q}_j\vec{f}(\omega')d\mu(\omega'),$$

where $\hat{X}_0(\omega,\omega')$ is a generalized operator function, and $\chi_{ij}(\omega,\omega')$ being the characteristic function of the set Ω_{ij}. The proof of this statement consists of the formal calculation of the operator $\hat{A}\hat{X} - \hat{X}\hat{N}$.

§4 Choice of Parameters

Until now we have not imposed any constraints on the measurable space $(\Omega, \mathfrak{A}, \partial\mu)$ used to extend the initial algebra $C^\infty(B(H_0))$ to the algebra $C^\infty(B(H))$; no constraints have been imposed on the projection operator $\hat{P}(\hat{P}(H) = H_0)$ and on other operators involved in the formulas defining $\hat{\Gamma}$. They play the role of parameters whose choice is at our disposal. To back up the formal results of the previous section, we need to choose these parameters in such a manner that the right-hand side of Eq. (3.3.5) will correctly define bounded operators in H.

The simplest way to choose the measurable set is to take for Ω an arbitrary compact of the complex plane C and for $d\mu$ an arbitrary finite Borel measure concentrated on this compact. After such a choice, the extended space

$$H = L^2_\mu(\Omega, H_0)$$

consists of measurable vector functions $\vec{f}(z)$ mapping the compact Ω into the space H_0, while the scalar product in H is defined by the equality

$$(\vec{f}(z), \vec{g}(z))_H = \mu^{-1} \int_{\Omega} (\vec{f}(z), \vec{g}(z))_{H_0} d\mu(z) \qquad (\mu = \mu(\Omega)).$$

According to Eq. (3.1.2), for a projection operator onto the subspace $H_0 \subset H$, one may take every operator of the form

$$\hat{P}(\vec{f}(z)) = \int \hat{p}(z)(\vec{f}(z)) d\mu(z), \tag{3.4.1}$$

where

$$\hat{p}(z) \in L^2_\mu(\Omega, B(H_0)), \qquad \int \hat{p}(z) d\mu(z) = I.$$

The choice of the remaining parameters is closely related to the estimates of the singular integrals that enter the right-hand side of the formal equalities (3.3.5). Accurate estimates of such integrals were obtained by Carleson, Muckenhoupt, and others. The reader can find detailed presentations of these questions in the survey [3] and in the monograph [20], both of which contain comprehensive bibliographies. We shall now give the results we need in the form suitable for our aims. The proofs are omitted.

1) Link lengths of the rectified Jordan curve generate a measure on it that is denoted by $|dz|$. Henceforth, we shall consider only such curves whose sets of self-intersection points have the length equal to zero. These curves, as well as the sets of points in the complex plane that make up the curves, will always be denoted by the same letter $\mathcal{L}$. The measure $|dz|$ on the curve $\mathcal{L}$ induces a measure on the complex plane concentrated on the set $\mathcal{L}$. It is also denoted by $|dz|$. A curve $\mathcal{L}$ is said to be a Carleson curve if its part lying in an arbitrary circle with the radius r has the length not exceeding Cr, where C is constant. The Hilbert space of functions $f(z)(z \in C)$ with the scalar product

$$(f, g) = \int_{\mathcal{L}} f(z)\overline{g(z)} |dz|$$

is denoted by $L^2(\mathcal{L})$.

A nonnegative function $w(z)$ is said to be the Muckenhoupt weight if

$$\sup_{z' \in \mathcal{L}} \sup_{r>0} \left\{ \frac{1}{r} \int_{\mathcal{L} \cap B(z', r)} w(z) |dz| \cdot \frac{1}{r} \int_{\mathcal{L} \cap B(z', r)} w(z)^{-1} |dz| \right\} < \infty,$$

where $B(z', r) = \{z : |z - z'| < r\}$.

The set of all Muckenhoupt weights on the curve $\mathcal{L}$ is denoted by $A_2(\mathcal{L})$. Here and henceforth all the sets and functions involved are assumed to be measurable with respect to the measure introduced above; principal value of singular integrals will be denoted by $\text{V}\!\!\int$.

MUCKENHOUPT–DAVID THEOREM. *If $\mathcal{L}$ is a Carleson curve and $w(z) \in A_2(\mathcal{L})$, then the operator*

$$\widehat{K}(f(z)) = \fint_{\mathcal{L}} \sqrt{\frac{w(z)}{w(\xi)}}\, \frac{f(\xi)}{\xi - z}|d\xi| = \lim_{\epsilon \to 0} \int_{|\xi - z| > \epsilon} \sqrt{\frac{w(z)}{w(\xi)}}\, \frac{f(\xi)}{\xi - z}|d\xi|$$

is bounded in the space $L^2(\mathcal{L})$.

2) Introduce the following notation:
$C^+(C^-)$—closed upper (lower) half-plane;
Ω—support of the measure $d\mu(z)$; $\Omega^+ = \Omega \cap C^+, \Omega^- = \Omega \setminus \Omega^+$;
μ^+, μ^-—restriction of the measure μ onto the sets Ω^+, Ω^-;
$L^2_\mu, L^2_{\mu^+}, L^2_{\mu^-}, L^2$—Hilbert spaces of functions with such scalar products:

$$(f, g)_\mu = \int f(z)\overline{g(z)}d\mu(z),$$
$$(f, g)_{\mu^\pm} = \int f(z)\overline{g(z)}d\mu^\pm(z),$$
$$(f, g) = \int_{-\infty}^{\infty} f(x)\overline{g(x)}dx;$$

$B(H_1, H_2)$ — set of all linear bounded operators mapping the space H_1 into the space H_2;
$\square(I)$ — square whose one diagonal is a segment I of the real axis;
$|I|$ — length of the segment I.

Let $w(x) \in A_2(-\infty, \infty)$ be an arbitrary Muckenhoupt weight on the real axis. The measure $d\mu(z)$ is said to be a $w(x)$–Carleson one if the inequality

$$\mu(\square(I)) \leqslant C \int_I w(x)dx$$

holds for all squares $\square(I)$. The set of all $w(x)$–Carleson measures is denoted by $K(w(x))$.

GENERALIZED CARLESON THEOREM. *Let*

$$w(x) \in A_2(-\infty, \infty), \qquad \mu \in K(w(x)), \qquad \nu \in K\left(w(x)^{-1}\right),$$

and let the operators $\widehat{D}_1, \widehat{D}_2, \widehat{D}_3^\pm$ be defined by the equalities

$$\widehat{D}_1(f) = \fint_{-\infty}^{\infty} \frac{f(x)}{\sqrt{w(x)}(x - z)}dx, \qquad \widehat{D}_2(f) = \sqrt{w(x)} \fint \frac{g(z')}{z' - x}d\nu(z'),$$
$$\widehat{D}_3^\pm(f) = \int \frac{g(z')}{z' - \bar{z}}d\nu^\pm(z').$$

Then

$$\widehat{D}_1 \in B(L^2, L^2_\mu), \qquad \widehat{D}_2 \in B(L^2_\nu, L^2), \qquad \widehat{D}_3^\pm \in B(L^2_{\nu^\pm}, L^2_{\mu^\pm}).$$

The measure $d\mu(z)$ is said to satisfy the $w(x)$–Carleson condition in the point $a \in (-\infty, \infty)$ if the inequality

$$\mu(\square(I)) \leqslant C \int_I w(x)\,dx$$

holds for all squares $\square(I)$ with the centre in point a. Note that for measures $d\mu(x)$ defined on the real axis, this inequality assumes for the form

$$\mu([a-h, a+h]) \leqslant C \int_{a-h}^{a+h} w(x)\,dx.$$

Consider an arbitrary finite system of disjoint intervals $\Delta_k = (a_k, b_k)$ positioned symmetrically with respect to the coordinate origin ($a_k = -b_{-k}, -N \leqslant k \leqslant N$), and denote by $\chi_G(x), \chi_F(x)$ characteristic functions of the sets $G = \cup_{k=-N}^{N} \Delta_k, F = R^1 \backslash G$. The following lemma is a simple corollary of the Carleson theorem.

LEMMA 3.4.1. *Let $w(x) \in A_2(-\infty, \infty)$ and the measure $d\mu(x)$ given on the real axis satisfy the $w(-x)$–Carleson condition on the ends of the intervals Δ_k:*

$$\mu([\pm a_k - h, \pm a_k + h]) \leqslant C \int_{\pm a_k - h}^{\pm a_k + h} w(-x)\,dx$$
$$= C \int_{\mp a_k - h}^{\mp a_k + h} w(x)\,dx \quad (-N \leqslant k \leqslant N).$$

Then the operators

$$\widehat{A}(f) = \int_{-\infty}^{\infty} \frac{\chi_F(x)\chi_G(y)}{x+y} \frac{f(x)}{\sqrt{w(x)}}\,dx,$$
$$\widehat{C}(g) = \int_{-\infty}^{\infty} \frac{\chi_F(x)\chi_G(y)}{x+y} \frac{g(y)}{\sqrt{w(x)}}\,d\mu(y)$$

are bounded:

$$\widehat{A} \in B(L^2, L^2_\mu), \qquad \widehat{C} \in B(L^2_\mu, L^2).$$

PROOF: Since the function

$$u(y) = \widehat{A}(f) = \int_{-\infty}^{\infty} \frac{\chi_F(x)\chi_G(y)}{\sqrt{w(x)}(x+y)} f(x)\,dx$$

differs from zero only in the intervals (a_k, b_k),

$$u(y) = \sum_{k=-N}^{N} \chi_k(y)\left[u_k(y) + v_k(y)\right],$$

where $\chi_k(y)$ is a characteristic function of the interval (a_k, b_k), while

$$u_k(y) = \int_{-a_k}^{\infty} \frac{\chi_F(x) f(x)}{\sqrt{w(x)}(x+y)} dx,$$
$$v_k(y) = \int_{-\infty}^{-b_k} \frac{\chi_F(x) f(x)}{\sqrt{w(x)}(x+y)} dx.$$

Therefore,

$$\begin{aligned}
\|\widehat{A}(f)\|^2_{L^2_\mu} &= \int_{-\infty}^{\infty} |u(y)|^2 d\mu(y) \\
&= \sum_{k=-N}^{N} \int_{a_k}^{b_k} |u_k(y) + v_k(y)|^2 d\mu(y) \\
&\leqslant 2 \sum_{k=-N}^{N} \int_{a_k}^{b_k} \left(|u_k(y)|^2 + |v_k(y)|^2\right) d\mu(y) \\
&\leqslant 2 \sum_{k=-N}^{N} \left(\int_{a_k}^{\infty} |u_k(y)|^2 d\mu(y) + \int_{-\infty}^{b_k} |v_k(y)|^2 d\mu(y) \right) \\
&= 2 \sum_{k=-N}^{N} \left(\int_{0}^{\infty} |u_k(a_k+\xi)|^2 d\mu(a_k+\xi) + \int_{-\infty}^{0} |v_k(b_k+\xi)|^2 d\mu(b_k+\xi \right)
\end{aligned}$$

and, if the functions $u_k(a_k+\xi), v_k(b_k+\xi)$ satisfy the inequalities

$$\begin{aligned}
&\int_0^{\infty} |u_k(a_k+\xi)|^2 d\mu(a_k+\xi) \leqslant C_k \|f|^2_{L^2}, \\
&\int_{-\infty}^{0} |v_k(b_k+\xi)|^2 d\mu(b_k+\xi) \leqslant C'_k \|f\|^2_{L^2},
\end{aligned} \tag{3.4.2}$$

then

$$\|\widehat{A}(f)\|^2_{L^2_\mu} \leqslant 2 \sum_{k=-N}^{N} (C_k + C'_k)\|f\|^2_{L^2}.$$

So in order to prove the lemma, it is sufficient to show the validity of inequality (3.4.2). Both inequalities are proved in similar fashion, so we shall prove only the first one.

Consider the auxiliary function

$$\varphi_k(z) = \int_{-a_k}^{\infty} \frac{|f(x)|}{\sqrt{w(x)}(x-z)} dx \qquad (\operatorname{Im} z > 0)$$

and the measure $d\mu_k(z)$ defined by the equality

$$\int \psi(z) d\mu_k(z) = \int_0^{\infty} \psi(-a_k + i\xi) d\mu(a_k+\xi).$$

This measure is obviously concentrated on a ray $-a_k + i\xi$ $(0 \leqslant \xi \leqslant \infty)$ and, for any square $\Box(I)$ with the vertex on this ray, holds the following inequality

$$\mu_k(\Box(I)) = \mu\left([a_k, a_k + \frac{1}{2}|I|]\right) \leqslant C \int\limits_{a_k - \frac{1}{2}|I|}^{a_k + \frac{1}{2}|I|} w(-x)dx$$

$$= C \int\limits_{-a_k - \frac{1}{2}|I|}^{-a_k + \frac{1}{2}|I|} w(x)dx = C \int\limits_I w(x)dx.$$

That is why $\mu_k \in K(w(x))$ and, in accordance with the generalized Carleson theorem,

$$\int |\varphi_k(z)|^2 d\mu_k(z) = \int_0^\infty |\varphi_k(-a_k + i\xi)|^2 d\mu(a_k + \xi) \leqslant C_k \|f\|^2_{L^2}.$$

Note now, that the definitions of the functions $u_k(y), \varphi_k(z)$ imply the following inequalities $(\xi > 0)$:

$$|u_k(a_k + \xi)| \leqslant \int_{-a_k}^\infty \frac{|f(x)|}{\sqrt{w(x)}(x + a_k + \xi)} dx \leqslant \int_{-a_k}^\infty \frac{|f(x)|}{\sqrt{w(x)[(x + a_k)^2 + \xi^2]}} dx,$$

$$\varphi_k(-a_k + i\xi) = \varphi_k^{(1)}(\xi) + i\varphi_k^{(2)}(\xi),$$

where

$$\varphi_k^{(1)}(\xi) = \int_{-a_k}^\infty \frac{|f(x)|(x + a_k)}{\sqrt{w(x)}[(x + a_k)^2 + \xi^2]} dx$$

$$\geqslant \frac{1}{\sqrt{2}} \int_{-a_k + \xi}^\infty \frac{|f(x)|}{\sqrt{w(x)[(x + a_k)^2 + \xi^2]}} dx,$$

$$\varphi_k^{(2)}(\xi) = \int_{-a_k}^\infty \frac{|f(x)|\xi}{\sqrt{w(x)}[(x + a_k)^2 + \xi^2]} dx$$

$$\geqslant \frac{1}{\sqrt{2}} \int_{-a_k}^{-a_k + \xi} \frac{|f(x)|}{\sqrt{w(x)[(x + a_k)^2 + \xi^2]}} dx.$$

Therefore,

$$2|\varphi_k(-a_k + i\xi)|^2 = 2\left(\varphi_k^{(1)}(\xi)^2 + \varphi_k^{(2)}(\xi)^2\right)$$

$$\geqslant \left(\varphi_k^{(1)}(\xi) + \varphi G_k^{(2)}(\xi)\right)^2 \geqslant \frac{1}{2}|u_k(a_k + \xi)|^2$$

and, hence

$$\int_0^\infty |u_k(a_k + \xi)|^2 d\mu(a_k + \xi) \leqslant 4 \int_0^\infty |\varphi_k(-a_k + i\xi)|^2 d\mu(a_k + \xi)$$

$$\leqslant 4C_k \|f\|^2_{L^2}.$$

It would be interesting to find a generalization of this lemma for an arbitrary open set $\mathcal{G}$, symmetric with respect to the coordinate origin.

3) every one-to-one correspondence $z \to \pi(z)$ of the complex plane is related to a measure transformation $\mu \to \mu\pi$ which determines the $\mu\pi$ -measure of the set Δ as a μ -measure of the set $\pi^{-1}(\Delta) : \mu\pi(\Delta) = \mu\left(\pi^{-1}(\Delta)\right)$. It is convenient to denote the measure μ by $d\mu(z)$ and the measure $\mu\pi$ by $d\mu\left(\pi(z)\right)$.

Evidently,

$$\int f\left(\pi(z)\right) d\mu\left(\pi(z)\right) \equiv \int f(z) d\mu(z) \quad \left(f(z) \in L^1_\mu\right),$$

and this identity can be taken as a definition of the transformation $\mu \to \mu\pi$. Note also that the set $\pi^{-1}(\Omega)$ is the support of the measure $\mu\pi$ (Ω is the support of the measure μ).

A measure $d\mu(z)$ is said to be π -invariant on the subset $\Omega_\pi \subset \Omega$ if $\pi(\Omega_\pi) = \Omega_\pi$ and $\mu\pi(\Delta) = \mu(\Delta)$ for all $\Delta \in \Omega_\pi$. The measure $d\mu(z)$ is π -invariant on Ω_π if and only if $\chi_\pi\left(\pi(z)\right) = \chi_\pi(z)$ and

$$\int \chi_\pi(z) f(z) d\mu(z) \equiv \int \chi_\pi(z) f\left(\pi(z)\right) d\mu(z)$$

for all functions $f(z) \in L^1_\mu$, where $\chi_\pi(z)$ is a characteristic function of the set Ω_π. For this reason we shall denote the π -invariant on the set Ω_π by the equality

$$\chi_\pi(z) d\mu(z) = \chi_\pi(z) d\mu\left(\pi(z)\right).$$

On the set of all vector functions $\vec{f}(z)$, the mapping $\pi(z)$ generates mutually invertible operators $\hat{\pi}, \hat{\pi}^{-1}$ defined by the equalities

$$\hat{\pi}\left(\vec{f}(z)\right) = \vec{f}\left(\pi(z)\right), \qquad \hat{\pi}^{-1}\left(\vec{f}(z)\right) = \vec{f}\left(\pi^{-1}(z)\right).$$

These operators implement isomorphism between the spaces

$$L^2_\mu(\Omega, H_0), L^2_{\mu\pi}\left(\pi^{-1}(\Omega), H_0\right),$$

and they commute with decomposable operators $\hat{a}(z)$ satisfying the condition $\hat{a}\left(\pi(z)\right) = \hat{a}(z)$.

If

$$\chi_\pi(z) d\mu(z) = \chi_\pi(z) d\mu\left(\pi(z)\right)$$

and

$$\vec{f}(z), \vec{g}(z) \in L^2_\mu(\Omega, H_0) = H,$$

then

$$\begin{aligned}
\left(\chi_\pi(z)\hat{\pi}(\vec{f}(z)), \vec{g}(z)\right)_H &= \mu^{-1} \int \left(\vec{f}(\pi(z)), \vec{g}(z)\right)_{H_0} \chi_\pi(z) d\mu(z) \\
&= \mu^{-1} \int \left(\vec{f}(z), \vec{g}(\pi^{-1}(z))\right) \chi_\pi(z) d\mu(z) \\
&= \left(\vec{f}(z), \chi_\pi(z)\hat{\pi}^{-1}(\vec{g}(z))\right)_H,
\end{aligned}$$

whence we have that the operator $\chi_\pi(z)\hat{\pi}$ is bounded in the space H, and the equality

$$(\chi_\pi(z)\hat{\pi})^* = \chi_\pi(z)\hat{\pi}^{-1} \tag{3.4.3}$$

is valid. In particular, if $\Omega_\pi = \Omega$, then the operator $\hat{\pi}$ is unitary in the space H.

If the mapping $\pi(z)$ is isometric on the Carleson curve $\mathcal{L}(|\pi(z_1) - \pi(z_2)| = |z_1 - z_2|$ for all $z_1, z_2 \in \mathcal{L})$, then the measure $|dz|$ generated by this curve is obviously π -invariant, and the operator $\hat{\pi}$ is unitary in the space $L^2(\mathcal{L})$.

Hence, in accordance with the Muckenhoupt–David theorem, the operator $\hat{\pi}\widehat{K}$:

$$\hat{\pi}\widehat{K}(f) = \int \sqrt{\frac{w(\pi(z))}{w(\xi)}} \frac{f(\xi)}{\xi - \pi(z)} d\xi$$

is also bounded in the space $L^2(\mathcal{L})$.

Let us denote by π_1, π_3, the mappings $\pi_1(z) = -z, \pi_3(z) = -\bar{z}$. It is clear that these equations transform $w(x)$–Carleson measures μ into $w(-x)$–Carleson measures $\mu\pi_1, \mu\pi_3$:

$$\mu \in K(w(x)) \Rightarrow \mu\pi_1, \mu\pi_3 \in K(w(-x))\,.$$

Since the operators $\hat{\pi}_1, \hat{\pi}_3$ isomorphically map the space $L^2_\mu(\Omega)$ into the spaces $L^2_{\mu\pi_1}(\pi_1^{-1}(\Omega)), L^2_{\mu\pi_3}(\pi_3^{-1}(\Omega))$, then, by the generalized Carleson theorem,

$$\hat{\pi}_1\widehat{D}_1 \in B(L^2, L^2_\mu), \qquad \hat{\pi}_1\widehat{D}_2 \in B(L^2_\nu L^2), \qquad \hat{\pi}_3\widehat{D}_3^\pm \in B(L^2_{\nu\pm}, L^2_{\mu\pm})$$

if $\mu \in K(w(-x))$ and $\nu \in K(w(x)^{-1})$.

Therefore, if $\mu = \nu \in K(w(-x)) \cap K(w(x)^{-1})$ and

$$\hat{\pi}_1\widehat{D}_1(f) = \int_{-\infty}^{\infty} \frac{f(x)}{\sqrt{w(x)}(x+z)} dx,$$

$$\hat{\pi}_1\widehat{D}_2(g) = \int \frac{\sqrt{w(-x)}g(z')}{z'+x} d\mu(z'),$$

$$\hat{\pi}_3\widehat{D}_3^\pm(g) = \int \frac{g(z')}{z'+z} d\mu^\pm(z'),$$

then

$$\hat{\pi}_1\widehat{D}_1 \in B(L^2, L^2_\mu), \qquad \hat{\pi}_1\widehat{D}_2 \in B(L^2_\mu, L^2), \qquad \hat{\pi}_3\widehat{D}_3^\pm \in B(L^2_{\mu\pm}, L^2_{\mu\pm}). \tag{3.4.4}$$

Note that the definition of the $w(z)$–Carleson measure, the generalized Carleson theorem, and its corollaries may be extended on the Muckenhoupt weights $w(z) \in A_2(|z| = 1)$ given on the unit circle $|z| = 1$.

Prior to considering concrete equations, two general remarks are in order.

It would be advantageous to expand the definition domain of vector functions $\vec{f}(z) \in L^2_\mu(\Omega, H_0)$ to the entire complex plane, assuming $\vec{f}(z) = 0$ when $z \notin \Omega$. This convention would result in simplifying the notation.

In the set of solutions of the same nonlinear equation, it will be convenient to introduce the equivalence relation, assuming $\hat{u} \sim \hat{v}$ if there exists such an invertible operator $\widehat{C} \in B(H)$ that $\hat{u} = \widehat{C}^{-1}\hat{v}\widehat{C}$. Indeed, substitution of $\hat{v}$ for $\hat{u}$ can be treated just as a change of a basis in the space H.

If the operator functions $\widehat{\Gamma}_1, \widehat{\Gamma}_2 \in C^\infty(B(H))$ are related by the equality $\widehat{\Gamma}_1 = \widehat{D}\widehat{\Gamma}_2\widehat{C}$, with the operators $\widehat{D}_1\widehat{C} \in B(H)$ being invertible, then their logarithmic derivatives $\hat{\gamma}_i = \widehat{\Gamma}_i^{-1}\delta\widehat{\Gamma}_i$ are homothetic ($\hat{\gamma}_1 = \widehat{C}^{-1}\hat{\gamma}_2\widehat{C}$) and therefore are equivalent solutions of the same nonlinear equation. This enables us to simplify, without loss of generality, the structure of the operators $\widehat{\Gamma}$ by multiplying them on both sides by appropriate operators.

KdV equation:

According to formulas (3.2.II(1)) and (3.3.II(1)), the operators $\widehat{\Gamma}$ have the form

$$\widehat{\Gamma} = e^{\hat{\theta}}\widehat{C}_1 + e^{-\hat{\theta}}\widehat{C}_2, \qquad \hat{\theta} = \hat{a}\left(x - (\hat{a}^2 + \lambda)t\right), \qquad (\lambda \in R^1),$$

and an additional condition $\partial\widehat{\Gamma}(I - \widehat{P}) = \widehat{\Gamma}\widehat{N}(I - \widehat{P})$ is reduced to the equations

$$(\hat{a}\widehat{C}_1 - \widehat{C}_1\widehat{N})(I - \widehat{P}) = 0, \qquad (\hat{a}\widehat{C}_2 + \widehat{C}_2\widehat{N})(I - \widehat{P}) = 0$$

in the coefficients $\widehat{C}_1, \widehat{C}_2 \in B(H)$.

The main case is when one of the operators $\widehat{C}_1, \widehat{C}_2$ (for example, $\widehat{C}_1$) is invertible. Then, without loss of generality, one can take the operator $\widehat{\Gamma}\widehat{C}_1^{-1} = e^{\hat{\theta}} + e^{-\hat{\theta}}(\widehat{C}_2\widehat{C}_1^{-1})$ instead of $\widehat{\Gamma}$, that is, to set $\widehat{C}_1 = I$. To satisfy the additional equations in the case when $\widehat{C}_1 = I$, one must set $\widehat{N} = \hat{a}$. If, in the capacity of $\hat{a}$, a normal operator with a simple spectrum is taken, then, without loss of generality, it may be regarded as an operator of multiplication by the function iz. For this reason we take the operators $\hat{a}, \widehat{N}$ equal to the multiplication operator: $\hat{a} = \widehat{N} = iz$. After such a choice, the set $\{z' : \widehat{N}(z') = \hat{a}(z)\}$ ($\{z' : -\widehat{N}(z') = \hat{a}(z)\}$) consists of one point z $(-z)$, and, according to Eq. (3.3.5), the formal solutions of equations (3.3.II(1)) have such a form

$$\widehat{C}_1\left(\vec{f}(z)\right) = \hat{m}_1(z)\vec{f}(z) + \int \frac{\hat{r}_1(z)\widehat{P}(z')}{i(z'-z)}\,\vec{f}(z')d\mu(z'),$$

$$\widehat{C}_2\left(\vec{f}(z)\right) = \hat{m}_2(z)\vec{f}(-z) + \int \frac{\hat{r}_2(z)\widehat{P}(z')}{i(z'+z)}\,\vec{f}(z')d\mu(z').$$

Substituting these expressions into formula (3.2.II(1')), we obtain

$$\widehat{\Gamma}\left(\vec{f}(z)\right) = \hat{\mathcal{E}}(z)\left\{\hat{m}_1(z)\vec{f}(z) + \int \frac{\hat{r}_1(z)\widehat{P}(z')}{i(z'-z)}\vec{f}(z')d\mu(z')\right\}$$
$$+ \hat{\mathcal{E}}(-z)\left\{\hat{m}_2(z)\vec{f}(-z) + \int \frac{\hat{r}_2(z)\widehat{P}(z')}{i(z'+z)}\vec{f}(z')d\mu(z')\right\},$$

where

$$\hat{\mathcal{E}}(z) = \exp\{iz\left(x + (z^2 - \lambda)t\right)\} \qquad \lambda \in (-\infty, \infty).$$

In particular, setting $\hat{m}_1(z) = I$, $\hat{r}_1(z) = 0$, we obtain the following expression for the operators $\hat{\Gamma}$:

$$\hat{\Gamma} = \hat{\mathcal{E}}(z) + \hat{\mathcal{E}}(-z)\widehat{M} + \hat{\Lambda}), \tag{3.4.5}$$

where the operators $\widehat{M}, \hat{\Lambda}$ are formally defined by the equalities

$$\widehat{M}\left(\vec{f}(z)\right) = \hat{m}_2(z)\vec{f}(-z), \tag{3.4.5'}$$

$$\hat{\Lambda}\left(\vec{f}(z)\right) = \int \frac{\hat{r}_2(z)\widehat{P}(z')}{i(z'+z)}\vec{f}(z')d\mu(z'). \tag{3.4.5''}$$

As shown above, all the operators $\hat{\Gamma}$ are reduced to such a form if for $\widehat{C}_1$ we choose invertible operators and for $\hat{a}$ we choose normal operators with the simple spectrum.We confine ourselves solely to such operators.

Now let us determine the conditions under which the equalities (3.4.5′) and (3.4.5″) correctly define the bounded operators in the space $H = L^2_\mu(\Omega, H_0)$.

Let us introduce the sets

$$\Omega_1 = \{z : z \in \Omega, -z \notin \Omega\}, \qquad \Omega_2 = \{z : z \in \Omega, -z \in \Omega\}$$

and decompose the space $L^2_\mu(\Omega, H_0)$ into an orthogonal sum of its subspaces $L^2_\mu(\Omega_1, H_0)$, $L^2_\mu(\Omega_2 H_0)$. The operators of multiplication by the characteristic functions $\chi_1(z), \chi_2(z)$ of the sets Ω_1, Ω_2 serve obviously as orthogonal projectors $\hat{\chi}_1, \hat{\chi}_2$ onto the above subspaces.

We have assumed that $\vec{f}(z) = 0$ at $z \notin \Omega$, whence it follows that the operator $\widehat{M}$ can be represented in the form $\widehat{M} = \hat{m}_2(z)\chi_2(z)\hat{\pi}_1$ where $\hat{\pi}_1\left(\vec{f}(z)\right) = \vec{f}(-z)$. Therefore, if the measure $d\mu(z)$ on the set Ω_2 is even $(\chi_2(z)d\mu(z) = \chi_2(z)d\mu(-z))$ and $\hat{m}_2(z) \in L^\infty_\mu(\Omega, B(H_0))$, then this operator is bounded in the space $L^2_\mu(\Omega, H_0)$; besides, according to Eq. (3.4.4),

$$M^* = \chi_2(z)\hat{\pi}_1\hat{m}_2^*(z),$$

since $\hat{\pi}_1^{-1} = \hat{\pi}_1$.

The formal linear operation A defines in the space $H = L^2_\mu(\Omega_1 H_0) \oplus L^2_\mu(\Omega_2, H_0)$ a bounded operator

$$\widehat{A} = \hat{\chi}_1 A\hat{\chi}_1 + \hat{\chi}_1 A\hat{\chi}_2 + \hat{\chi}_2 A\hat{\chi}_1 + \hat{\chi}_2 A\hat{\chi}_2,$$

provided that each operation $\hat{\chi}_k A\hat{\chi}_l$ $(k, l = 1, 2)$ generates a bounded operator. Thus, Eq. (3.4.5″) defines a bounded operator $\hat{\Lambda}$ if the operations

$$\hat{\chi}_k\Lambda\hat{\chi}_l\left(\vec{f}(z)\right) = \int \frac{\chi_k(z)\hat{r}_2(z)\widehat{P}(z')\chi_l(z')}{i(z'+z)}\vec{f}(z^1)d\mu(z')$$

generate bounded operators.

We shall denote the distance between the sets $D_1, D_2 \subset C$ by $\operatorname{dist}(D_1, D_2)$.

The operation $\hat{\chi}_1\Lambda\hat{\chi}_2$ contains a singular integral, and properties of this operation are determined by the structure of the measure $d\mu(z)$ on the set Ω_2. Consider two extreme cases.

a) On the set Ω_2, the measure $d\mu(z)$ coincides with the plane Lebesgue measure: $d\mu(z) = dxdy$. It follows from the theory of the integral with the weak singularity (see, e.g., [18]) that the operation $\hat{\chi}_1\Lambda\hat{\chi}_2$ generates a bounded operator if

$$\chi_2(z)\hat{r}_2(z), \qquad \chi_2(z)\widehat{P}(z) \in L^\infty_\mu(\Omega, B(H_0)).$$

If $\operatorname{dist}(\Omega_1\Omega_2) > 0$, then the both sets Ω_1, Ω_2 are compact and there exists such a positive number d that for all the points $z_1, z_1' \in \Omega_1, z_2 \in \Omega_2$ the inequalities $|z_1, +z_1'| > d, |z_1 + z_2| > d$ are valid. Hence, $\chi_k(z)|z' + z|^{-1}\chi_l(z) \leqslant d^{-1}$ ($k \neq l$ or $k = l = 1$); from this we obtain an estimate for the norms of the operators:

$$\|\hat{\chi}_k\Lambda\hat{\chi}_l\|^2_H \leqslant (\mu d)^{-2}\left\{\int \chi_k(z)\|\hat{r}_2(z)\|^2_{H_0}d\mu(z)\int \chi_l(z')\|\widehat{P}(z')\|^2_{H_0}d\mu(z')\right\}.$$

Thus, Eq. (3.4.5″) defines a bounded operator $\hat{\Lambda}$ if the following conditions are met:

a_1) $d\mu(z) = dxdy$ on the set Ω_2;

a_2) $\operatorname{dist}(\Omega_1, \Omega_2) > 0$;

a_3) $\chi_2(z)\hat{r}_2(z), \quad \chi_2(z)\widehat{P}(z) \in L^\infty_\mu(\Omega, B(H_0))$;

a_4) $\chi_1(z)\hat{r}_2(z), \quad \chi_1(z)\widehat{P}(z) \in L^2_\mu(\Omega, B(H_0))$.

For the KdV equation, this case is but of slight interest, since having chosen the measure in the described way, one cannot prove the invertibility of the operators $\widehat{\Gamma}$ yielding real solutions for all real values x, t, and so the solutions obtained appear to be singular.

b) The set Ω_2 lies on the Carleson curve $\mathcal{L}$ which is symmetric with respect to the origin ($z \in \mathcal{L} \Rightarrow -z \in \mathcal{L}$), and the measure on it is defined as the arc length: $d\mu(z) = |dz|$. Since the mapping $\pi_1(z) = -z$ is isometric, then, according to the Muckenhoupt–David theorem, the operator

$$\hat{\pi}_1\widehat{K}(f(z)) = \int\sqrt{\frac{w(-z)}{w(z')}}\,\frac{f(z')}{i(z'+z)}\,|dz'|$$

is bounded in the space $L^2(\mathcal{L})$ if $w(z) \in A_2(\mathcal{L})$. Therefore, the operator $\hat{\chi}_2\hat{\pi}_1\widehat{K}\hat{\chi}_2$ is bounded in the space $L^2_\mu(\Omega)$ and induces the bounded operator in the space $H = L^2_\mu(\Omega, H_0)$. Multiplying it on both sides by the arbitrary decomposable operators $\hat{a}_1(z), \hat{a}_2(z) \in L^\infty_\mu(\Omega, B(H_0))$, we obtain the operator

$$\hat{a}_2\hat{\chi}_2\hat{\pi}_1\widehat{K}\hat{\chi}_2\hat{a}\left(\vec{f}(z)\right) = \int\frac{\chi_2(z)\hat{a}_2(z)\sqrt{w(-z)}\chi_2(z')a(z')}{i(z'+z)\sqrt{w(z')}}\,\vec{f}(z')d\mu(z')$$

bounded in the space H. Thus the operation $\hat{\chi}_2\Lambda\hat{\chi}_2$ generates a bounded operator in the space H, if b$_1$) $\Omega_2 \subset \mathcal{L}$ and $d\mu(z) = |dz|$ on Ω_2;

b$_2$) $\chi_2(z)\hat{f}_2(z)\,(w(-z))^{-\frac{1}{2}}, \chi_2(z)\widehat{P}(z)\,(w(z))^{\frac{1}{2}} \in L^\infty_\mu\,(\Omega, B(H_0))$, where $w(z) \in A_2(\mathcal{L})$.

As in the previous case, it follows from here that the entire operation Λ generates a bounded operator if, in addition,

b$_3$) $\operatorname{dist}(\Omega_1, \Omega_2) > 0$;

b$_4$) $\chi_1(z)\hat{f}_2(z), \quad \chi_1(z)\widehat{P}(z) \in L^2_\mu\,(\Omega, B(H_0))$.

The generalized Carleson theorem permits finding the criterion of boundedness of the operator $\hat{\Lambda}$ when $\operatorname{dist}(\Omega_1, \Omega_2) = 0$.

We shall consider the typical and most interesting case when the set Ω lies on the real and imaginary axes while its asymmetric part Ω_1 is contained in the finite number of intervals Δ_k on the imaginary axis $(ia_k^-, ia_k^+)(a_k^- < a_k^+)$ which are complementary to the set Ω_2.

Let us introduce the following notation:

$$\Omega_1^+ = \Omega_1 \cap C^+, \qquad \Omega_1^- = \Omega_1 \setminus \Omega_1^+, \qquad \Omega_2(R) = \Omega_2 \cap (-\infty, \infty),$$
$$\Omega_2(I) = \Omega_2 \setminus \Omega_2(R), \qquad \chi_1^+(z), \qquad \chi_1^-(z), \qquad \chi_2(R,z), \qquad \chi_2(I,z)$$

are characteristic functions of these sets, μ_1 is a restricton of the measure on the set Ω_1.

To make the operation Λ generate a bounded operator in the space H, it is sufficient to meet the conditions b$_1$), b$_2$), and

b_3') dist $(\Omega_1^+, \pi_1(\Omega_1^-)) > 0 \quad (\pi_1(z) = -z)$;

b_4') $\chi_1(z)\hat{f}_2(z), \quad \chi_1(z)\widehat{P}(z) \in L^\infty_\mu\,(\Omega, B(H_0))$;

b_5') at zero and at the ends of the intervals Δ_k, the measure μ_1 satisfies simultaneously both the $w(-z)$ and the $w(-z)^{-1}$ Carleson conditions, that is

$$\mu\left([-ih, ih]\right) \leqslant C \min\left\{\int_{-h}^{h} w(-x)dx, \int_{-h}^{h} w(-x)^{-1}dx\right\},$$

$$\mu\left([i(a_k^\pm - h), i(a_k^\pm + h)]\right) \leqslant C \min\left\{\int_{i(a_k^\pm - h)}^{i(a_k^\pm + h)} w(-z)|dz|, \int_{i(a_k^\pm - h)}^{i(a_k^\pm + h)} w(-z)^{-1}|dz|\right\}.$$

Indeed, conditions b_1 and b_2 imply the boundedness of the operator $\hat{\chi}_2\Lambda\hat{\chi}_2$ and, since

$$\hat{\chi}_1\Lambda\hat{\chi}_1 = \hat{\chi}_1^+\Lambda\hat{\chi}_1^+ + \hat{\chi}_1^+\Lambda\hat{\chi}_1^- + \hat{\chi}_1^-\hat{\Lambda}\hat{\chi}_1^+ + \hat{\chi}_1^-\Lambda\hat{\chi}_1^-$$
$$\hat{\chi}_1\Lambda\hat{\chi}_2 = \hat{\chi}_1\Lambda\hat{\chi}_2(R) + \hat{\chi}_1\Lambda\hat{\chi}_2(I), \qquad \hat{\chi}_2\Lambda\hat{\chi}_1 = \hat{\chi}_2(R)\Lambda\hat{\chi}_1 + \hat{\chi}_2(I)\Lambda\hat{\chi}_1,$$

it is sufficent to show that all the operators on the right-hand side of these equalities are bounded.

The condition b_3' implies existence of a number $d > 0$ with a property that $\hat{\chi}_1^-(z)|z+z'|^{-1}\hat{\chi}_1^+(z') \leqslant d^{-1}$ whence it follows that the operators $\hat{\chi}_1^+\Lambda\hat{\chi}_1^-, \hat{\chi}_1^-\Lambda\hat{\chi}_1^+$ are bounded if the condition b_4' is met. Further, the condition b_5' implies that

$\mu_1 \in K(w(-x)) \cap K(w(x)^{-1})$ and, according to the generalized Carleson theorem, the operators

$$\hat{\chi}_1^+ \Lambda \hat{\chi}_1^+, \qquad \hat{\chi}_1^- \Lambda \hat{\chi}_1^-, \qquad \hat{\chi}_1 \Lambda \hat{\chi}_2(R), \qquad \hat{\chi}_2(R) \Lambda \hat{\chi}_1$$

are bounded if the conditions b_2 and b_4' are met.

Finally, since the measure μ_1 at the ends of the intervals Δ_k simultaneously satisfies both the $w(-z)$ and the $w(-z)^{-1}$ Carleson conditions, the conditions b_2 and b_4', according to Lemma 3.4.1, imply that the operators $\hat{\chi}_1 \Lambda \hat{\chi}_2(I), \hat{\chi}_2(I) \Lambda \hat{\chi}_1$ are bounded.

Thus, the right-hand side of Eq. (3.4.5″) defines a bounded operator in the space $H = L\mu^2(\Omega, H_0)$ if either conditions b_1, b_2, b_3, b_4 or $b_1, b_2, b_3', b_4', b_5'$ are met. It goes without saying that these conditions are not necessary for the operator Λ to be bounded. It would be extremely interesting to find conditons that, when added to b_1 and b_2, form a set of necessary and sufficient conditions of the boundedness of the operator. The case when the set Ω lies on the real and imaginary axes is of special importance.

Toda lattice (TL) and Langmuir lattice (LL):

For these equations the operators $\hat{\Gamma}$ assume such a form

$$\hat{\Gamma}_s = \hat{a}^n e^{\hat{a}^s t} \hat{C}_1(s) + \hat{a}^{-n} e^{\hat{a}^{-s} t} \hat{C}_2,(s),$$

where $s = 1$ for TL and $s = 2$ for LL (see formulas (3.2.II(4)), (3.2.II(6))). Since we need only their logarithmic derivatives with respect to ∂_t and $\hat{\Gamma}_s^{-1} \partial_t \Gamma_s = (\hat{a}^{-n} \hat{\Gamma}_s)^{-1} \partial_t (\hat{a}^{-n} \hat{\Gamma}_s)$, these operators can be replaced by

$$\hat{\Gamma}_s = e^{\hat{a}^s t} \hat{C}_1(s) + \hat{a}^{-2n} e^{\hat{a}^{-s} t} \hat{C}_2(s).$$

In accordance with (3.3.II(4)), (3.3.II(6)), the additional conditions are reduced to the equations

$$\left(\hat{a}^s \hat{C}_1(s) - \hat{C}_1(s) \hat{N}(s)\right)(I - \hat{P}) = 0 \qquad (\hat{a}^{-s} \hat{C}_2(s) - \hat{C}_2(s) \hat{N}(s)(I - \hat{P}) = 0$$

in the coefficients $\hat{C}_1(s), \hat{C}_2(s) \in B(H)$.

In order to satisfy the former equation at $\hat{C}_1(s) = I$, one must take $\hat{N}(s) = \hat{a}^s$. In the capacity of $\hat{a}$, we shall take the operator of multiplication by a function, setting

$$\hat{a} = a(z) = (z - i)(z + i)^{-1}, \qquad \hat{N}(s) = a(z)^s, \qquad \hat{C}_1(s) = I.$$

Since $a(z)^{-1} = a(-z)$ and

$$\{z : a(z') = a(-z)\} = \{-z\}, \qquad \{z' : a(z')^2 = a(-z)^2\} = \{-z, z^{-1}\},$$

the formal solution of the latter equation, according to Eq. (3.3.5), has the form

$$\widehat{C}_2(1)\left(\vec{f}(z)\right) = \hat{m}_1(z)\vec{f}(-z) + \int \frac{\hat{r}_1(z)\hat{p}_1(z')}{a(-z)-a(z')}\vec{f}(z')d\mu^{(1)}(z'),$$

$$\widehat{C}_2(2)\left(\vec{f}(z)\right) = \hat{m}_2(z)\frac{\vec{f}(-z)+\vec{f}(z^{-1})}{2} + \hat{n}_2(z)\frac{\vec{f}(-z)-\vec{f}(z^{-1})}{2} + \int \frac{\hat{r}_2(z)\hat{p}_2(z')}{a(-z)^2-a(z')^2}\vec{f}(z')d\mu^{(2)}(z').$$

We choose the measure $\mu^{(2)}$ invariant with respect to the transformation $z \to -z^{-1}$ $(d\mu^{(2)}(z) = d\mu^{(2)}(-z^{-1}))$, and for the extended space for the operators $\widehat{\Gamma}_2$, we take the subspace of the space $L^2_{\mu^{(2)}}(\Omega, H_0)$ consisting of vector functions invariant with respect to this transformation $(\vec{f}(z) = \vec{f}(-z^{-1}))$. Since $a(-z^{-1}) = -a(z)$ and $a(-z^{-1})^2 = a(z)^2$, to make the operators $\widehat{\Gamma}_2, \hat{p}(z)$ map this space into itself, it is necessary to meet the following conditons:

$$\hat{p}(-z^{-1}) = \hat{p}(z), \qquad \hat{r}_2(-z^{-1}) = \hat{r}_2(z), \qquad \hat{m}_2(-z^{-1}) = \hat{m}_2(z).$$

The integral operator in the formula for $\widehat{C}_2(2)$ can be simplified. Indeed, since $a(-z^{-1}) = -a(z)$,

$$\frac{1}{a(-z)^2-a(\xi)^2} = \frac{1}{2a(-z)}\left\{\frac{1}{a(-z)-a(\xi)} + \frac{1}{a(-z)+a(\xi)}\right\} = \frac{1}{2a(-z)}\left\{\frac{1}{a(-z)-a(\xi)} + \frac{1}{a(-z)-a(-\xi^{-1})}\right\},$$

$$\int \frac{\hat{p}(\xi)\vec{f}(\xi)}{a(-z)-a(-\xi^{-1})}\,d\mu^{(2)}(\xi) = \int \frac{\hat{p}(-\eta^{-1})\vec{f}(-\eta^{-1})}{a(-z)-a(\eta)}\,d\mu^{(2)}(-\eta^{-1}) = \int \frac{\hat{p}(\eta)\vec{f}(\eta)}{a(-z)-a(\eta)}\,d\mu^{(2)}(\eta),$$

and

$$\int \frac{\hat{r}_2(z)\hat{p}(z')\vec{f}(z')}{a(-z)^2-a(z')^2}\,d\mu^{(2)}(z') = \frac{\hat{r}_2(z)}{a(-z)}\int \frac{\hat{p}(z')\vec{f}(z^1)}{a(-z)-a(z')}\,d\mu^{(2)}(z').$$

Thus, we arrive at the following formulas for the operators,

$$\widehat{\Gamma}_s = e^{a(z)^s t} + a(-z)^{2n} e^{a(-z)^s t}(\widehat{M}_s + \widehat{\Lambda}_s), \tag{3.4.6}$$

where $a(z) = (z-i)(z+i)^{-1}$,

$$\widehat{M}_s = \chi_2(z)\hat{m}_s(z)\hat{\pi}_1 \qquad (\pi_1(z) = -z) \tag{3.4.6'}$$

$$\widehat{\Lambda}_s\left(\vec{f}(z)\right) = \int \frac{a(z)^{s-1}\hat{r}_s(z)\hat{p}_s(z')}{a(-z)-a(z')}\,\vec{f}(z')d\mu^{(s)}(z') \tag{3.4.6''}$$

(with $d\mu^{(2)}(z) = d\mu^{(2)}(-z^{-1})$,

(3.4.7) $\hat{m}_2(z) = \hat{m}_2(-z^{-1}), \qquad \hat{p}_2(z) = \hat{p}_2(-z^{-1}), \qquad \hat{r}_2(z) = \hat{r}_2(-z^{-1}),$

and the operator $\widehat{\Gamma}_2$ acting in the space of the vector functions $\vec{f}(z) \in L^2_{\mu^{(2)}}(\Omega, H_0)$ that meet the condition $\vec{f}(z) = \vec{f}(-z^{-1})$).

To guarantee boundedness of the operators $\hat{a} = a(z), \hat{a}^{-1} = a(-z)$, it is necessary that the measure support Ω does not contain the points $\pm i$. Finally, the equality

$$\frac{1}{a(-z) - a(z')} = \frac{(z-i)(z'+1)}{2i(z'+z)}$$

implies that the operators $\widehat{M}_s, \widehat{\Lambda}_s$ coincide with the operators (3.4.5′) and (3.4.5″) whose conditions of boundedness have been found already.

N-wave equation and equation of chiral field theory:

In order to meet the additional conditions, the operator parameters must be chosen so that they satisfy the equations

$$(\hat{a}\widehat{C} - \widehat{C}\widehat{N})(I - \widehat{P}) = 0, \qquad [\hat{d}_1(x), \widehat{N}] = [\hat{d}_2(y), \widehat{N}] = 0$$

(see formulas (3.3.II(9)), (3.3.II(10))).

Let the operator functions $\hat{d}_1(x), \hat{d}_2(y)$ take their values from the commutative subalgebra $K(H_0)$ of the algebra $B(H_0)$. (For instance, $\hat{d}_1(x) = \widehat{D}_1 x, \hat{d}_2(y) = \widehat{D}_2 y$, where $\widehat{D}_1, \widehat{D}_2 \in B(H_0)$ and $[\widehat{D}_1, \widehat{D}_2] = 0$). Then the equations $[\hat{d}_1(x), \widehat{N}] = [\hat{d}_2(y), \widehat{N}] = 0$ are satisfied automatically if $\widehat{N}$ is the operator of multiplication by the function $N(z)$. So we replace $\widehat{N}$ by the operator of multiplication with $\bar{z}$ and $\hat{a}$ by the decomposable operator $\hat{a} = z\widehat{Q}_1(z) + \bar{z}\widehat{Q}_2(z)$, where $\widehat{Q}_1(z) \in L^2_\mu(\Omega, B(H_0)), \widehat{Q}_1(z)^2 = \widehat{Q}_1(z)$, and $\widehat{Q}_2(z) = I - \widehat{Q}_1(z)$. This choice permits to employ formula (3.3.5). Since in our case,

$$\Omega_{1,1} = \Omega_{1,2} = \{z' : \bar{z}' = z\} = \{\bar{z}\}, \qquad \Omega_{2,1} = \Omega_{2,2} = \{z' : \bar{z}' = \bar{z}\} = \{z\},$$

the formal solution of the equation $(\hat{a}\widehat{C} - \widehat{C}\widehat{N})(I - \widehat{P}) = 0$, according to (3.3.5), assumes the form:

$$\begin{aligned}\widehat{C}\left(\vec{f}(z)\right) = \widehat{Q}_1(z)&\left\{\overline{\chi}_2(z)\hat{m}(z)\hat{\pi}_2\left(\vec{f}(z)\right) + \hat{r}(z) \int \frac{\hat{p}(z')}{i(z-\bar{z}')}\vec{f}(z')d\mu(z')\right\} \\ &+ \widehat{Q}_2(z)\left\{\hat{m}(z)\left(\vec{f}(z)\right) + \hat{r}(z) \int \frac{\hat{p}(z')}{i(\bar{z}-z')}\vec{f}(z')d\mu(z')\right\}\end{aligned}$$

where $\overline{\chi}_2(z)$ is the characteristic function of the set $\overline{\Omega}_2 = \{z : z \in \Omega, \bar{z} \in \Omega\}$ and $\hat{\pi}_2(\vec{f}(z)) = \vec{f}(\bar{z})$.

Thus, we again arrive at the operators whose conditions of boundedness have already been found. We only need to substitute this expression into formulas (3.2.II(9′)) and (3.2.II(10)) that define the operators $\widehat{\Gamma}$. It is important that, first, the equalities

$$\widehat{Q}_i(z)^2 = \widehat{Q}_i(z), \qquad i = 1, 2, \qquad \widehat{Q}_1(z)\widehat{Q}_2(z) = 0$$

imply that

$$\hat{a}^n = z^n\widehat{Q}_1(z) + \bar{z}^n\widehat{Q}_2(z),$$
$$\widehat{A}_1^n = \left(-i(I+\hat{a})^{-1}\right)^n = (-i(1+z)^{-1})^n\widehat{Q}_1(z) + \left(-(i+\bar{z})^{-1}\right)^n \widehat{Q}_2(z),$$
$$\widehat{A}_2^n = \left(-i(I-\hat{a})^{-1}\right)^n = \left(-i(1-z)^{-1}\right)^n \widehat{Q}_1(z) + \left(-i(1-\bar{z})^{-1}\right)^n \widehat{Q}_2(z)$$

and, second, that the operators of multiplication by scalar functions commute with decomposable operators, while the operator functions $\hat{d}_1(x), \hat{d}_2(y)$ commute with the operator $\hat{\pi}_2$ and with the operator function $\hat{p}(z')$, provided that its values belong to the same subalgebra $K(H_0)$.

This permits, after substituting the expression found for the operators $\widehat{C}$ into the series (3.2.II(9′)) and (3.2.II(10)), to carry out summation of the series. It will result in such expressions for the operators $\widehat{\Gamma}$:

$$\begin{aligned}\widehat{\Gamma}\left(\vec{f}(z)\right) = \widehat{Q}_1(z)&\Big\{\bar{\chi}_2(z)\hat{m}(z)\widehat{\mathcal{E}}(z)\hat{\pi}_2\left(\vec{f}(z)\right)\\ &+\hat{r}(z)\widehat{\mathcal{E}}(z)\int\frac{\hat{p}(z')}{i(z-\bar{z}')}\vec{f}(z')d\mu(z')\Big\}\\ &+\widehat{Q}_2(z)\Big\{\hat{m}(z)\widehat{\mathcal{E}}(\bar{z})\vec{f}(z)\\ &+\hat{r}(z)\widehat{\mathcal{E}}(\bar{z})\int\frac{\hat{p}(z')}{i(\bar{z}-\bar{z}')}\vec{f}(z')d\mu(z')\Big\},\end{aligned} \tag{3.4.8}$$

where

$$\widehat{\mathcal{E}}(z) = \exp\left\{iz\left(\hat{d}_1(x)+\hat{d}_2(y)\right)\right\} \qquad \text{N-wave equation} \tag{3.4.8'}$$

$$\widehat{\mathcal{E}}(z) = \exp i\left\{\frac{\hat{d}_2(y)}{z-1} - \frac{\hat{d}_1(x)}{z+1}\right\}, \qquad \text{(chiral field equation)},$$

while the operator functions $\hat{d}_1(x), \hat{d}_2(y), \hat{p}(z')$ take their values from the same commutative subalgebra $K(H_0) \subset B(H_0)$. (In the N-wave equation, the operator functions $\hat{d}_1(x), \hat{d}_2(y)$ are replaced, for convenience sake, by $i\hat{d}_1(x), id_2(y)$).

Let the measure by concentrated in the upper closed half-plane and let $\chi_0(z), \chi_+(z)$ be characteristic functions of the sets formed by intersection of the support Ω of the measure $d\mu(z)$ with the real axis and the upper open half-plane, respectively. We shall choose the measure $d\mu(z)$ and the operators $\widehat{Q}_2(z), \hat{r}(z), \hat{m}(z)$, so that the equalities

$$\chi_0(z)d\mu(z) = \chi_0(z)\frac{1}{2\pi}|dz|, \qquad \hat{m}(z) = I,$$
$$\chi_0(z)Q_2(z) = \chi_+(z)\widehat{Q}_2(z)\hat{r}(z) = 0, \qquad \chi_0(z)\hat{r}(z) = \hat{r}_0(z)\hat{q}(z),$$
$$\chi_+(z)\hat{r}(z) = \hat{q}(z), \qquad \left[\hat{q}(z), \widehat{\mathcal{E}}(z)\right] = 0$$

be valid. Then

$$\chi_0(z)\widehat{Q}_1(z) = \chi_0(z)\left[I - \widehat{Q}_2(z)\right] = \chi_0(z),$$
$$\bar{\chi}_2(z) = \chi_0(z), \qquad \bar{\chi}_2(z)\hat{\pi}_2 = \chi_0(z)\hat{\pi}_2 = \chi_0(z),$$

and formula (3.4.8) will assume the form

$$((3.4.8'')) \quad \widehat{\Gamma}\left(\vec{f}(z)\right) = \left\{\chi_0(z)\widehat{\mathcal{E}}(z) + \chi_+(z)\widehat{Q}_2(z)\widehat{\mathcal{E}}(\bar{z})\right\}\left(\vec{f}(z)\right) + \left\{\chi_0(z)\hat{r}_0(z)\widehat{\mathcal{E}}(z) + \chi_+(z)\widehat{Q}_1(z)\widehat{\mathcal{E}}(z)\right\} \int \frac{\hat{q}(z)\hat{p}(z')}{i(z-\bar{z}')}\vec{f}(z')d\mu(z').$$

Modified KdV equation (MKdV), nonlinear Schrödinger–Heisenberg equation (NSH), sine-Gordon equation (SG):

For these equations it follows from the formulas 3.2.II(2)), (3.2.II(3′)), (3.2.II(7′)) and the corresponding formulas of §3 that

$$\widehat{\Gamma} = e^{\hat{\theta}}\widehat{C}\widehat{Q}_1 + e^{-\hat{\theta}}\widehat{C}\widehat{Q}_2,$$

$$(3.4.9) \qquad \widehat{Q}_1 = \frac{1}{2}(I+\widehat{B}), \qquad \widehat{Q}_2 = \frac{1}{2}(I-\widehat{B}), \qquad \widehat{B}^2 = I,$$

while the additional condition is expressed by the equation

$$(3.4.9') \qquad (\hat{a}\widehat{C} - \widehat{C}\widehat{B}\widehat{N})(I-\widehat{P}) = 0$$

in $\widehat{C}$. Thus the operators $\widehat{\Gamma}$ differ only by the form of the operator function $\hat{\theta}$ that is equal to $\hat{a}x - \hat{a}^3t$ for the MKdV equation, to $\hat{a}x - \varepsilon^{-1}\hat{a}^2t$ for the NSH equation, and to $\hat{a}x + \hat{a}^{-1}t$ for the SG equation.

We confine our consideration to the most interesting case when the operator $\widehat{B} = \widehat{B}^* \neq \pm I$ is self-adjoint and $\varepsilon = -i$. Here the operators $\widehat{Q}_1, \widehat{Q}_2$ are orthogonal projectors onto the subspaces $H_1 = \widehat{Q}_1(H), H_2 = \widehat{Q}_2(H)$ whose orthogonal sum yields the entire space $H = H_1 \oplus H_2$. In accordance with this decomposition, we shall write the elements $\vec{f}(z) \in H$ as a column,

$$\vec{f}(z) = \begin{pmatrix} \vec{f}_1(z) \\ \vec{f}_2(z) \end{pmatrix}, \qquad \vec{f}_1(z) \in H_1, \qquad \vec{f}_2(z) \in H_2,$$

and the operators $\widehat{T} \in B(H)$ as operator matrices

$$\widehat{T} = \begin{pmatrix} \widehat{T}_{11} & \widehat{T}_{12} \\ \widehat{T}_{21} & \widehat{T}_{22} \end{pmatrix}, \qquad \widehat{T}_{ij} \in B(H_j, H_i).$$

The operator $\widehat{T}$ commutes with $\widehat{B}$ if and only if its matrix is a diagonal one ($\widehat{T}_{12} = \widehat{T}_{21} = 0$), for instance,

$$\widehat{Q}_1 = \begin{pmatrix} I_1 & 0 \\ 0 & 0 \end{pmatrix}, \qquad \widehat{Q}_2 = \begin{pmatrix} 0 & 0 \\ 0 & I_2 \end{pmatrix}, \qquad B = \begin{pmatrix} I_1 & 0 \\ 0 & -I_2 \end{pmatrix},$$

where I_i is a unit operator in the space H_i. Since the projector $\widehat{P}$ must commute with the operator $\widehat{B}$, the decomposable operator $\hat{p}(z)$ in formula (3.4.1) must be taken in the form of a diagonal matrix

$$\hat{p}(z) = \begin{pmatrix} \hat{p}_1(z) & 0 \\ 0 & \hat{p}_2(z) \end{pmatrix}, \qquad \hat{p}_i(z) \in L^2_\mu(\Omega, B(H_i)), \qquad \int \hat{p}_i(z)d\mu(z) = I_i.$$

We shall take the operators $\hat{a}$ and $\widehat{N}$ also in the form of diagonal matrices, setting

$$\hat{a} = a_1(z)\widehat{Q}_1 + a_2(z)\widehat{Q}_2, \qquad \widehat{N} = n_1(z)\widehat{Q}_1 - n_2(z)\widehat{Q}_2,$$
$$a_1(z) = n_1(z) = iz, \qquad a_2(z) = n_2(z) = iz.$$

Then $\widehat{B}\widehat{N} = n_1(z)\widehat{Q}_1 + n_2(z)\widehat{Q}_2$ and the sets $\Omega_{ij} = \{z' : n_j(z') = a_i(z)\}$ associated with Eq. (3.4.9′) contain one point each:

$$\Omega_{11} = \Omega_{22} = \{z\}, \qquad \Omega_{12} = \Omega_{21} = \{\bar{z}\}.$$

Whence, according to formula (3.3.5), we find that the formal solution of this equation, written down in the matrix form $\widehat{C} = (\widehat{C}_{ij})$, assumes the form

(3.4.10′)
$$\widehat{C}_{11}\left(\vec{f}_1(z)\right) = \hat{m}_{11}(z)\vec{f}_1(z) + \int \frac{\hat{r}_{11}(z)\hat{p}_1(z')}{i(z'-z)}\vec{f}_1(z')d\mu(z')$$
$$\widehat{C}_{12}\left(\vec{f}_2(z)\right) = \hat{m}_{12}(z)\vec{f}_2(\bar{z}) + \int \frac{\hat{r}_{12}(z)\hat{p}_2(z')}{i(\bar{z}'-z)}\vec{f}_2(z')d\mu(z')$$

(3.4.10″)
$$\widehat{C}_{21}\left(\vec{f}_1(z)\right) = \hat{m}_{21}(z)\vec{f}_1(\bar{z}) + \int \frac{\hat{r}_{21}(z)\hat{p}_1(z')}{i(z'-\bar{z})}\vec{f}_1(z')d\mu(z')$$
$$\widehat{C}_{22}\left(\vec{f}_2(z)\right) = \hat{m}_{22}(z)\vec{f}_2(z) + \int \frac{\hat{r}_{22}(z)\hat{p}_2(z')}{i(\bar{z}'-\bar{z})}\vec{f}_2(z')d\mu(z').$$

Substituting these expressions into formula (3.4.9), we express $\widehat{\Gamma}$ as follows,

$$\widehat{\Gamma} = \widehat{\mathcal{E}}\begin{pmatrix}\widehat{C}_{11} & 0\\ 0 & \widehat{C}_{22}\end{pmatrix} + \widehat{\mathcal{E}}^{-1}\begin{pmatrix}0 & \widehat{C}_{12}\\ \widehat{C}_{21} & 0\end{pmatrix}, \qquad \widehat{\mathcal{E}} = \widehat{\mathcal{E}}(z) = \begin{pmatrix}e^{\theta(z)} & 0\\ 0 & e^{\overline{\theta(z)}}\end{pmatrix},$$

where

(3.4.11)
$$\theta(z) = iz(x+z^2t) \qquad \text{for MKdV}$$
$$\theta(z) = iz(x-zt) \qquad \text{for NSH}$$
$$\theta(z) = i(zx - z^{-1}t) \qquad \text{for SG}$$

In particular, if $\widehat{C}_{11} = I_1, \widehat{C}_{22} = I_2$, then

(3.4.12)
$$\widehat{\Gamma} = \widehat{\mathcal{E}} + \widehat{\mathcal{E}}^{-1}\begin{pmatrix}0 & \widehat{C}_{12}\\ \widehat{C}_{21} & 0\end{pmatrix},$$

where the operators $\widehat{C}_{12}, \widehat{C}_{21}$ are formally determined by the right-hand sides of Eqs. (3.4.10′) and (3.4.10″).

Various criteria of boundedness of these operators are derived exactly as it has been done for the KdV equation. We shall give one criterion.

Let us decompose the support Ω of the measure μ into a sum of disjoint subsets

$$\Omega_1 = \{z : z \in \Omega, \bar{z} \notin \Omega\}, \qquad \Omega_2 = \{z : z \in \Omega, \bar{z} \in \Omega\};$$

each of the subsets will be partitioned, in their turn, into the following parts:

$$\Omega_i^{\pm} = \Omega_i \cap C^{\pm}, \qquad i = 1,2 \qquad \Omega_2^0 = \Omega_2 \cap (-\infty, \infty).$$

The characteristic functions for these sets will be denoted by

$$\chi_1(z), \chi_2(z), \chi_1^{\pm}(z), \chi_2^{\pm}(z), \chi_0(z),$$

while the restriction of the measure $d\mu(z)$ on the set with the characteristic function $\chi(z)$ will be denoted by $\chi(z)d\mu(z)$.

The operators $\widehat{C}_{12}, \widehat{C}_{21}$ determined by the right-hand side of Eqs. (3.4.10′), (3.4.10″) are bounded in the space $H = L^2_{\mu}(\Omega, H_0)$ if

1) the set Ω_2 lies on the Carleson curve $\mathcal{L}$ and $\chi_2(z)d\mu(z) = \chi_2(z)|dz|$;
2) $\operatorname{dist}(\Omega_1, \Omega_2^+ \cup \Omega_2^-) > 0, \qquad \operatorname{dist}(\Omega_1^+), \pi_2(\Omega_1^-)) > 0$;
3) $\chi_1(z)d\mu(z) \in K(w(x)) \cap K(w(x)^{-1})$;
4) $\hat{m}_2(z), \quad \chi_2(z)\hat{r}_2(z)(w(\bar{z}))^{-\frac{1}{2}}, \quad \chi_2(z)\hat{p}(z)(w(z))^{\frac{1}{2}},$
$\chi_1(z)\hat{r}_2(z), \quad \chi_1(z)\hat{p}(z) \in L^{\infty}_{\mu}(\Omega, B(H_0))$.

Here $w(z)$ stands for an arbitrary Muckenhoupt weight on the curve $\mathcal{L}$, while the condition (3) accounts for the restriction of this weight on the real axis.

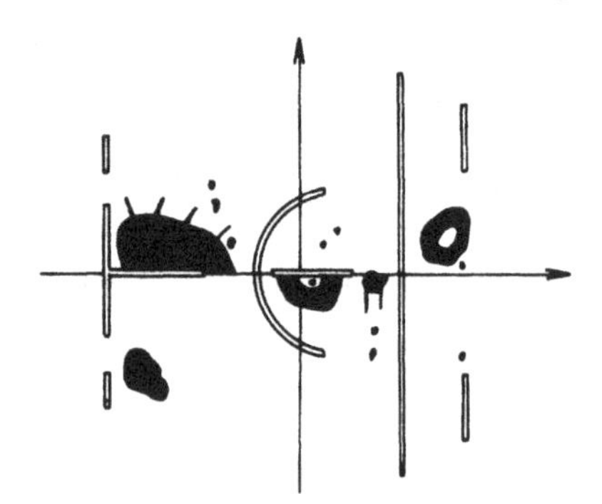

Figure 1

Fig. 1 shows a typical example of the set Ω meeting the listed conditions. Curves making up the subset Ω_2 are shown with double lines, while the subset Ω_1 is represented by points, and the curves and domains by black.

§5 Properties of Logarithmic Derivatives with Respect to the Conjugating Operation

In the previous section, we showed how to choose the measure and operator parameters to ensure that the solutions $\widehat{\Gamma}$ of the linear equations in the left-hand part of Table II would belong to the algebra $C^{\infty}(B(H))$ and satisfy the additional conditions that allow to apply the operation of projection onto the original algebra $C^{\infty}(B(H_0))$. Now we must find the conditions when the operators $\widehat{\Gamma}$ are invertible. Besides, we must take care to ensure that the solutions of nonlinear equations obtained via the projection operation possess certain additional properties. For example, for the physical applications, only real solutions of the KdV, KP, or sine-Gordon equations are of interest. Solutions of other nonlinear equations must also satisfy some limitations to be of physical interest. Therefore it would be convenient to find the conditions which would simultaneously satisfy these limitations and meet the invertibility condition.

We will proceed as follows. In this section we assume that the operators $\widehat{\Gamma}$ are invertible and find the conditions to be met by the measure and operator parameters of the solutions obtained for satisfying the appropriate limitations, and in the subsequent section we shall find the invertibility conditions for the operators $\widehat{\Gamma}$ contained in this section.

Now we shall view some standard notation: Ω is a support of the measure μ; $\chi_2(z), \bar{\chi}_2(z)$ are characteristic functions of the sets

$$\Omega_2 = \{z : z \in \Omega, -z \in \Omega\}, \qquad \bar{\Omega}_2 = \{z : z \in \Omega, \bar{z} \in \Omega\};$$

and $\hat{\pi}_1, \hat{\pi}_2, \hat{\pi}_3, \hat{\pi}_4$ are the operators generated by the mappings

$$\pi_1(z) = -z, \qquad \pi_2(z) = \bar{z}, \qquad \pi_3(z) = -\bar{z}, \qquad \pi_4(z) = -z^{-1}.$$

The equality $d\mu(z) = d\mu(\pi(z))$ means that the measure is π-invariant on the entire support, while the equality $\chi_\pi(z)d\mu(z) = \chi_\pi(z)d\mu(\pi(z))$ means that it is π -invariant on the subset $\Omega_\pi = \pi(\Omega_\pi) \subset \Omega$ whose characteristic function is $\chi_\pi(z)$. The limitations to be imposed on the solutions of nonlinear equations are described by certain relations between them and operator functions conjugate to them. That is why, first of all, we have to elucidate the properties of logarithmic derivatives of the operators $\widehat{\Gamma}$ with respect to the conjugating operation.

First, let us consider the operators (3.4.5) whose logarithmic derivatives $\hat{\gamma} = \widehat{\Gamma}^{-1}\partial_x\widehat{\Gamma}$ (and also the operators $\widehat{P}\hat{\gamma}\widehat{P}$) solve the KdV equations in the algebra $C^\infty(B(H))$ (and in the algebra $C^\infty(B(H_0))$, respectively). By their construction, the operators $\widehat{C}_2 = \widehat{\mathrm{M}} + \widehat{\Lambda}$ satisfy the equation $\hat{a}\widehat{C}_2 + \widehat{C}_2\hat{a} = \hat{r}_2(z)\widehat{P}$, where $\hat{a} = iz$. Hence, $\partial_x\widehat{\Gamma} = \widehat{\Gamma}iz - \mathcal{E}(-z)\hat{r}_2(z)\widehat{P}$, which implies

$$\hat{\gamma} = iz - \widehat{\Gamma}^{-1}\mathcal{E}(-z)\hat{r}_2(z)\widehat{P}.$$

Let us transform this formula and Eq. (3.4.5) into a more symmetric form. To this end, we denote the common left multiplier of the operators $\hat{r}_2(z), \hat{m}_2(z)$ by $\hat{r}(z)$:

$$\hat{r}_2(z) = \hat{r}(z)\hat{q}(z), \qquad \hat{m}_2(z) = \hat{r}(z)\hat{m}(z)$$

and take out the operator $\mathcal{E}(z) = \mathcal{E}(-z)^{-1}$ to the left of the parentheses. As a result, we arrive at the following expressions for the operators $\widehat{\Gamma}$ and its logarithmic derivative:

$$\left.\begin{aligned} \widehat{\Gamma} = \mathcal{E}(z)\widehat{T}, \qquad \widehat{T} = I + \widehat{R}(z)\widehat{L}, \\ \hat{\gamma} = iz - \widehat{T}^{-1}\widehat{R}(z)\hat{q}(z)\widehat{P} \end{aligned}\right\}, \tag{3.5.1}$$

where

$$\begin{gathered} \mathcal{E}(z) = \exp iz\left\{x + (z^2 - \lambda)t\right\}, \\ \widehat{R}(z) = \mathcal{E}(-z)^2\hat{r}(z) = \mathcal{E}(-z)\hat{r}(z)\mathcal{E}(-z), \\ \widehat{L}\left(\vec{f}(z)\right) = \chi_2(z)\hat{m}(z)\hat{\pi}_1\left(\vec{f}(z)\right) + \int\frac{\hat{q}(z)\hat{p}(z')}{i(z+z')}\vec{f}(z')d\mu(z'). \end{gathered} \tag{3.5.1'}$$

Likewise, for the operators $\widehat{\Gamma}$ defined by equalities (3.4.6), (3.4.12) and their logarithmic derivatives, we arrive at the following expressions:

$$\widehat{\Gamma}=\widehat{\mathcal{E}}(z)\widehat{T},\qquad \widehat{T}=I+\widehat{R}(z)\widehat{L}, \tag{3.5.2}$$

$$\hat{\gamma}_s=\widehat{\Gamma}^{-1}\partial_t\widehat{\Gamma}=a(z)^s+\widehat{T}^{-1}\widehat{R}(z)\hat{q}(z)a(-z)^{s-1}\widehat{P}\qquad (s=1,2)$$

$$\hat{\gamma}=\widehat{\Gamma}^{-1}\partial_x\widehat{\Gamma}=\widehat{N}(z)-\widehat{T}^{-1}\widehat{R}(z)\hat{q}(z)\widehat{B}\widehat{P}, \tag{3.5.3}$$

where for $TL(s=1)$ and $LL(s=2)$,

$$\begin{aligned} R(z)&=\mathcal{E}_s(z)^{-1}\hat{r}_s(z)\mathcal{E}_s(-z),\qquad \hat{q}(z)=\hat{q}_s(z),\\ \mathcal{E}_s(z)&=a(z)^n\exp a(z)^s t,\qquad a(z)=(z-i)(z+i)^{-1},\\ \widehat{L}\left(\vec{f}(z)\right)&=\widehat{L}_s\left(\vec{f}(z)\right)=\chi_2(z)\hat{m}_s(z)\hat{\pi}_1\left(\vec{f}(z)\right)\\ &\quad+\int\frac{\hat{q}_s(z)\hat{p}_s(z')}{a(-z)-a(z')}\vec{f}(z')d\mu_s(z'), \end{aligned} \tag{3.5.2'}$$

while for the MKdV and NSH equations

$$\begin{aligned} &\widehat{R}(z)=\widehat{\mathcal{E}}(z)^{-1}\hat{r}(z)\widehat{\mathcal{E}}^*(z)^{-1},\qquad \hat{r}(z)=\widehat{Q}_1\hat{r}(z)\widehat{Q}_2+\widehat{Q}_2\hat{r}(z)\widehat{Q}_1,\\ &\widehat{\mathcal{E}}(z)=\exp\left\{\theta(z)\widehat{Q}_1+\overline{\theta(z)}\widehat{Q}_2\right\},\\ &\widehat{N}(z)=iz\widehat{Q}_1+\overline{iz}\widehat{Q}_2,\\ &\widehat{L}\left(\vec{f}(z)\right)=\bar{\chi}_2(z)\hat{m}(z)\hat{\pi}_2\left(\vec{f}(z)\right)\\ &\qquad+\int\hat{q}(z)\left\{\frac{\widehat{Q}_1}{i(z'-\bar{z})}+\frac{\widehat{Q}_2}{i(\bar{z}'-z)}\right\}\hat{p}(z')\vec{f}(z')d\mu(z'),\\ &\left[\hat{m}(z),\widehat{Q}_i\right]=\left[\hat{q}(z),\widehat{Q}_i\right]=\left[\hat{p}(z),\widehat{Q}_i\right]=0,\qquad (i=1,2), \end{aligned} \tag{3.5.3'}$$

the function $\theta(z)$ being determined by equalities (3.4.11). (In the above calculations, we took into account that, for the operators $\hat{r}=\widehat{Q}_1\hat{r}\widehat{Q}_2+\widehat{Q}_2\hat{r}\widehat{Q}_1$ and $A=c\widehat{Q}_1+\bar{c}\widehat{Q}_2$ $(c\in\mathbf{C})$, the commutative relation $\widehat{A}\hat{r}=\hat{r}\widehat{A}^*$ holds which, in particular, implies the equality

$$\widehat{R}(z)=\widehat{\mathcal{E}}(z)^{-2}\hat{r}(z)=\widehat{\mathcal{E}}(z)^{-1}\hat{r}(z)\widehat{\mathcal{E}}^*(z)^{-1}).$$

The operators $\widehat{\Gamma}$ defined by formula (3.4.12) solve the equation $\partial_x\widehat{\Gamma}=\widehat{N}(z)\widehat{B}\widehat{\Gamma}\widehat{B}$, whence, according to Eq. (3.5.3),

$$\widehat{\Gamma}^{-1}\widehat{N}(z)\widehat{B}\widehat{\Gamma}\widehat{B}=\widehat{N}(z)-\widehat{\Gamma}^{-1}\widehat{\mathcal{E}}(z)\widehat{R}(z)\hat{q}(z)\widehat{B}\widehat{P}.$$

If the operator $\widehat{N}(z)$ is invertible (i.e., the measure support contains no zero), then the logarithmic derivative $\hat{\gamma}=\widehat{\Gamma}^{-1}\partial_x\widehat{\Gamma}=\widehat{\Gamma}^{-1}\widehat{N}(z)\widehat{B}\widehat{\Gamma}\widehat{B}$ is invertible as well, and $\hat{\gamma}^{-1} = \widehat{B}\widehat{\Gamma}^{-1}\widehat{B}\widehat{N}(z)^{-1}\widehat{\Gamma}$. Multiplying both parts of the previous equality by the operator $\widehat{B}\widehat{\Gamma}^{-1}\widehat{B}\widehat{N}(z)^{-1}\widehat{\Gamma}$, we find that

$$\begin{aligned} I &= \hat{\gamma}^{-1}\widehat{N}(z) - \widehat{B}\widehat{\Gamma}^{-1}\widehat{B}\widehat{N}(z)^{-1}\widehat{\mathcal{E}}(z)\widehat{R}(z)\hat{q}(z)\widehat{B}\widehat{P} \\ &= \hat{\gamma}^{-1}\widehat{N}(z) - \widehat{B}\widehat{T}^{-1}\widehat{\mathcal{E}}(z)^{-1}\widehat{B}\widehat{N}(z)^{-1}\widehat{\mathcal{E}}(z)\widehat{R}(z)\hat{q}(z)\widehat{B}\widehat{P} \\ &= \hat{\gamma}^{-1}\widehat{N}(z) + \widehat{B}\widehat{T}^{-1}\widehat{R}(z)\widehat{N}^*(z)^{-1}\hat{q}(z)\widehat{P}, \end{aligned}$$

since

$$\widehat{R}(z)\widehat{B} = -\widehat{B}\widehat{R}(z), \qquad \widehat{N}(z)^{-1}\widehat{R}(z) = \widehat{R}(z)\widehat{N}^*(z)^{-1},$$

and the operators $\widehat{N}(z), \widehat{B}, \widehat{\mathcal{E}}(z), \hat{q}(z)$ are commutative. Combining these equalities and formulas (3.5.3), we see that when the operator $\widehat{N}(z)$ is invertible, the mutually inverse operators $\widehat{N}(z)^{-1}\hat{\gamma}$ and $\hat{\gamma}^{-1}\widehat{N}(z)$ have the form

$$\begin{aligned} \widehat{N}(z)^{-1}\hat{\gamma} &= I - \widehat{N}(z)^{-1}\widehat{T}^{-1}\widehat{R}(z)\widehat{B}\hat{q}(z)\widehat{P}, \\ \hat{\gamma}^{-1}\widehat{N}(z) &= I - \widehat{B}\widehat{T}^{-1}\widehat{R}(z)\widehat{N}^*(z)^{-1}\hat{q}(z)\widehat{P}. \end{aligned} \tag{3.5.3''}$$

Let the operators $\hat{d}_1(x), \hat{d}_2(y)$ in formula (3.4.8″) be self-adjoint. Then $\mathcal{E}(z)^{-1} = \mathcal{E}(\bar{z})^*$, which allows us to represent the operator $\widehat{\Gamma}$ in the form:

$$\widehat{\Gamma} = \widehat{\mathcal{E}}(z)\widehat{T}_1, \qquad \widehat{T}_1 = \widehat{C}(z) + \widehat{R}(z)\widehat{L}, \tag{3.5.4}$$

where the operators $\widehat{\mathcal{E}}(z)$ are determined by the equalities (3.4.8′), and

$$\begin{aligned} \widehat{C}(z) &= \chi_0(z) + \chi_+(z)\widehat{\mathcal{E}}(\bar{z})^*\widehat{Q}_2(z)\widehat{\mathcal{E}}(\bar{z}), \\ \widehat{R}(z) &= \chi_0(z)\widehat{\mathcal{E}}(z)^*\hat{r}_0(z)\widehat{\mathcal{E}}(z) + \chi_+(z)\widehat{\mathcal{E}}(\bar{z})^*\widehat{Q}_1(z)\widehat{\mathcal{E}}(z), \\ \widehat{L}\left(\vec{f}(z)\right) &= \int\frac{\hat{q}(z)\hat{p}(z')}{i(\bar{z}'-z)}\vec{f}(z')d\mu(z'). \end{aligned} \tag{3.5.4'}$$

Having calculated the logarithmic derivatives $\hat{\gamma}_1 = \widehat{\Gamma}^{-1}\partial_x\widehat{\Gamma}$, $\hat{\gamma}_2 = \widehat{\Gamma}^{-1}\partial_y\widehat{\Gamma}$, we find that for the N-wave problem,

$$\hat{\gamma}_1\widehat{D}_1^{-1} = \hat{\gamma}_2\widehat{D}_2^{-1} = i\bar{z} - \widehat{T}^{-1}\widehat{R}(z)\hat{q}(z)\widehat{P}, \tag{3.5.5}$$

while for the chiral field equations,

$$\hat{\gamma}_\alpha A_\alpha(\bar{z})^{-1}\widehat{D}_\alpha^{-1} = I + (-1)^\alpha\widehat{T}_1^{-1}\widehat{R}(z)A_\alpha(z)\hat{q}(z)\widehat{P} \quad (\alpha = 1, 2), \tag{3.5.6}$$

where

$$\begin{aligned} A_1(z) &= i(z-1)^{-1}, & A_2(z) &= -i(z+1)^{-1}, \\ \widehat{D}_1 &= \partial_x\left(\hat{d}_1(x)\right), & \widehat{D}_2 &= \partial_y\left(\hat{d}_2(y)\right). \end{aligned}$$

For invertible operators, we arrive at the following formulas

$$\widehat{D}_\alpha A_\alpha(\bar{z})\hat{\gamma}_\alpha^{-1} = I - (-1)^\alpha A_\alpha(\bar{z})\widehat{T}_1^{-1}\widehat{R}(z)\hat{q}(z)\widehat{P}, \qquad (\alpha = 1, 2), \tag{3.5.6'}$$

whose proof is quite similar to the one given above for formulas (3.5.3″).

Let us clarify now the properties which the operators $\widehat{L}$ posses with respect to the conjugating operation. If the operators

$$\widehat{A}_1\left(\vec{f}(z)\right) = \int \frac{\hat{q}(z)\hat{p}(z')}{i(z'-\bar{z})}\vec{f}(z')d\mu(z'),$$
$$\widehat{A}_2\left(\vec{f}(z)\right) = \int \frac{\hat{q}(z)\hat{p}(z')}{i(\bar{z}'-z)}\vec{f}(z')d\mu(z')$$

are bounded in the space $L^2_\mu(\Omega, H_0)$, then, by definition,

$$\widehat{A}_1^*\left(\vec{f}(z)\right) = \int \frac{\hat{p}^*(z)\hat{q}^*(z')}{i(z'-\bar{z})}\vec{f}(z')d\mu(z'),$$
$$\widehat{A}_2^*\left(\vec{f}(z)\right) = \int \frac{\hat{p}^*(z)\hat{q}^*(z')}{i(\bar{z}'-z)}\vec{f}(z')d\mu(z'),$$

whence it follows that these operators are self-adjoint, provided that $\hat{q}(z) = \hat{p}^*(z)$:

$$\hat{q}(z) = \hat{p}^*(z) \Rightarrow \widehat{A}_1 = \widehat{A}_1^*, \qquad \widehat{A}_2 = \widehat{A}_2^*. \tag{3.5.7}$$

If $a(z) = (z-i)(z+i)^{-1}$, and the operator

$$\widehat{C}\left(\vec{f}(z)\right) = \int \frac{\hat{q}(z)\hat{p}(z')}{a(\bar{z}) - a(z')}\vec{f}(z')d\mu(z')$$

is bounded, then

$$\widehat{C}^*\left(\vec{f}(z)\right) = \int \frac{\hat{p}^*(z)\hat{q}^*(z')}{\overline{a(\bar{z}')} - \overline{a(z)}}\vec{f}(z')d\mu(z')$$

and, since

$$\overline{a(\bar{z}')} - \overline{a(z)} = a(z')^{-1} - a(\bar{z})^{-1} = a(z')^{-1}a(\bar{z})^{-1}\left(a(\bar{z}) - a(z')\right),$$

we have

$$\widehat{C}^*\left(\vec{f}(z)\right) = \int \frac{a(\bar{z})\hat{p}^*(z)a(z')\hat{q}^*(z')}{a(\bar{z}) - a(z')}\vec{f}(z')d\mu(z').$$

So we see that the condition for the operator $\widehat{C}$ to be self-adjoint is the validity of the quality $\hat{q}(z) = a(\bar{z})\hat{p}^*(z) = (a(-z)\hat{p}(z))^*$:

$$\hat{q}(z) = a(\bar{z})\hat{p}^*(z) \Rightarrow \widehat{C} = \widehat{C}^*. \tag{3.5.8}$$

If in formulas (3.5.1′) and (3.5.2′), the measure $d\mu(z)$ is π_3-invariant ($d\mu(z) = d\mu(\pi_3(z))$), then the operator $\hat{\pi}_3 = \hat{\pi}_3^{-1}$ is bounded in the space $H = L^2_\mu(\Omega, H_0)$, which enables us to carry out the following substitutions in these formulas: $\widehat{L} = \hat{\pi}_3^2\widehat{L} = \hat{\pi}_3\widehat{L}_1$, $\widehat{L}_1 = \hat{\pi}_3\widehat{L}$, whence

$$\widehat{L}_1\left(\vec{f}(z)\right) = \chi_2(z)\hat{\pi}_3\hat{m}(z)\hat{\pi}_1\left(\vec{f}(z)\right) + \int\frac{\hat{q}(-\bar{z})\hat{p}(z')}{i(z'-\bar{z})}\vec{f}(z')d\mu(z'),$$

$$\widehat{L}_{1,s}\left(\vec{f}(z)\right) = \chi_2(z)\hat{\pi}_3\hat{m}_s(z)\hat{\pi}_1\left(\vec{f}(z)\right) + \int\frac{\hat{q}_s(-\bar{z})\hat{p}_s(z')}{a(\bar{z})-a(z')}\vec{f}(z')d\mu(z').$$

According to Eqs. (3.5.7) and (3.5.8), the integral operators on the right-hand side of these equalities will be self-adjoint, if

$$\hat{q}(z) = \hat{p}^*(-\bar{z}), \qquad \hat{q}_s(z) = a(-z)\hat{p}_s^*(-\bar{z}) = \left(a(\bar{z})\hat{p}_s(-\bar{z})\right)^*.$$

Further, from the definition of the operators $\hat{\pi}$ and formula (3.4.3) it follows, that if the measure satisfies the condition $\chi_\pi(z)d\mu(z) = \chi_\pi(z)d\mu\left(\pi(z)\right)$, then

$$\pi\chi_\pi(z)\hat{m}(z) = \chi_\pi(z)\hat{m}\left(\pi(z)\right)\hat{\pi},$$
$$\left(\chi_\pi(z)\hat{m}(z)\hat{\pi}\right)^* = \chi_\pi(z)\hat{m}^*\left(\pi^{-1}(z)\right)\hat{\pi}^{-1}.$$

In particular, $\hat{\pi}_2 = \hat{\pi}_2^{-1}$ implies

$$\left(\bar{\chi}_2(z)\hat{m}(z)\hat{\pi}_2\right)^* = \bar{\chi}_2(z)\hat{m}(\bar{z})\hat{\pi}_2$$

if

$$\bar{\chi}_2(z)d\mu(z) = \bar{\chi}_2(z)d\mu(\bar{z}).$$

Thus, if the measure $d\mu(z)$ and the decomposable operator $\hat{m}(z)$ meet the conditions

$$\bar{\chi}_2(z)d\mu(z) = \bar{\chi}_2(z)d\mu(\bar{z}), \qquad \hat{m}(z) = \hat{m}(\bar{z})^*,$$

then the operator $\bar{\chi}_2(z)\hat{m}(z)\hat{\pi}_2$ is self-adjoint.

Since a π_3-invariant measure $d\mu(z)$ satisfying the condition $\chi_2(z)d\mu(z) = \chi_2(z)d\mu(-z)$ meets the condition $\bar{\chi}_2(z)d\mu(z) = \bar{\chi}_2(z)d\mu(\bar{z})$ automatically, and $\hat{\pi}_3\hat{\pi}_1 = \hat{\pi}_2$, then the operator

$$\hat{\pi}_3\chi_2(z)\hat{m}(z)\hat{\pi}_1 = \chi_2(z)\hat{m}(-\bar{z})\hat{\pi}_3\hat{\pi}_1 = \chi_2(z)\hat{m}(-\bar{z})\hat{\pi}_2$$

will be self-adjoint if

$$d\mu(z) = d\mu(-\bar{z}), \qquad \chi_2(z)d\mu(z) = \chi_2(z)d\mu(-z), \qquad \hat{m}(z) = \hat{m}^*(\bar{z}).$$

Note, finally, that the operators $\hat{\pi}$ transform constant vector functions into themselves: if $\vec{f} \in H_0$, then $\hat{\pi}\vec{f} = \vec{f}$. Thus the equalities $\hat{\pi}\widehat{P} = \widehat{P}$ hold for the projectors $\widehat{P}$ onto the space H_0; from these equalities it follows, for one, that

$$\hat{\pi}_3\hat{q}(z)\widehat{P} = \hat{q}(-\bar{z})\hat{\pi}_3\widehat{P} = \hat{q}(-\bar{z})\widehat{P}.$$

The above considerations can be summed up in the following.

LEMMA 3.5.1. *If in formulas (3.5.3′) and (3.5.4′) the measures and decomposable operators $\hat{m}(z), \hat{q}(z)$ satisfy the conditions*

$$\bar{\chi}_2(z)d\mu(z) = \bar{\chi}_2(z)d\mu(\bar{z}), \qquad \hat{m}(z) = \hat{m}^*(\bar{z}), \qquad \hat{q}(z) = \hat{p}^*(z),$$

then the operators $\widehat{L}$ and $\bar{\chi}_2(z)\hat{\pi}_2$ are self-adjoint, and

$$\widehat{T}^{-1}\widehat{R}(z)\hat{q}(z) = \widehat{T}^{-1}\widehat{R}(z)\hat{p}^*(z), \qquad \widehat{T}_1^{-1}\widehat{R}(z)\hat{q}(z) = \widehat{T}_1^{-1}\widehat{R}(z)\hat{p}^*(z).$$

If in formulas (3.5.1′) and (3.5.2′) the measures and decomposable operators $\hat{m}(z), \hat{q}(z), \hat{q}_s(z)$ satisfy the conditions

$$d\mu(z) = d\mu(-\bar{z}), \qquad \chi_2(z)d\mu(z) = \chi_2(z)d\mu(-z),$$
$$\hat{m}(z) = \hat{m}^*(\bar{z}), \qquad \hat{q}(z) = \hat{p}^*(-\bar{z}), \qquad \hat{q}_s(z) = a(-z)\hat{p}_s^*(-\bar{z}),$$

then the corresponding operators T can be represented in the form $\widehat{T} = I + \widehat{R}(z)\hat{\pi}_3\widehat{L}_1$, with $\widehat{L}_1 = \widehat{L}_1^*$, and

$$\widehat{T}^{-1}\widehat{R}(z)\hat{q}(z)\widehat{P} = \widehat{T}^{-1}\widehat{R}(z)\hat{\pi}_3\hat{p}^*(z)\widehat{P},$$
$$\widehat{T}^{-1}\widehat{R}_s(z)\hat{q}_s(z)a(-z)^{s-1}\widehat{P} = \widehat{T}^{-1}\widehat{R}_s(z)\hat{\pi}_3 a(\bar{z})^s\hat{p}^*(z)\widehat{P}.$$

Now it is time to show how the projection operation affects the properties of solutions obtained for nonlinear equations. Note, first of all, that by applying the projection operation to the operator $\widehat{A} \in B(H)$, we obtain the operator $\widehat{P}\widehat{A}\widehat{P}$, which is identified in a natural way with the corresponding operator from the algebra $B(H_0)$, also denoted by $\widehat{P}\widehat{A}\widehat{P}$. For this reason and to avoid ambiguity, we stipulate that imbedding $\widehat{P}\widehat{A}\widehat{P} \in B(H_0)$ (and $\widehat{P}\widehat{A}\widehat{P} \in C^\infty(B(H_0))$ respectively), would mean that this operator is considered in the space H_0.

LEMMA 3.5.2. *Let*

$$\widehat{A} \in B(H), \qquad \hat{p}(z) \in L^2_\mu(\Omega, B(H_0)),$$
$$\widehat{P}\left(\vec{f}(z)\right) = \int \hat{p}(z)\vec{f}(z)d\mu(z)$$

and

$$\int \hat{p}(z)d\mu(z) = I.$$

Then the operator $\hat{v}^* \in B(H_0)$, conjugated to the operator $\hat{v} = \widehat{P}\widehat{A}\hat{p}^*(z)\widehat{P} \in B(H_0)$, is obtained by substituting $\widehat{A}$ for $\widehat{A}^*$, that is

$$\left(\widehat{P}\widehat{A}\hat{p}^*(z)\widehat{P}\right)^* = \widehat{P}\widehat{A}^*\hat{p}^*(z)\widehat{P}.$$

PROOF: If $\vec{f}, \vec{g} \in H_0$, then $\widehat{P}\vec{f} = \vec{f}$, $\widehat{P}, \vec{g} = \vec{g}$, and

$$\hat{v}(\vec{f}) = \widehat{P}\widehat{A}\hat{p}^*(z)\widehat{P}(\vec{f}) = \widehat{P}\widehat{A}\hat{p}^*(z)(\vec{f}) = \int \hat{p}(z')\widehat{A}\hat{p}^*(z)\vec{f}d\mu(z'),$$

whence we get such a chain of equalities,

$$\begin{aligned}
(\hat{v}\vec{f},\vec{g})_{H_0} &= \left(\int \hat{p}(z')\widehat{A}\hat{p}(z)\vec{f}d\mu(z'),\vec{g}\right)_{H_0} \\
&= \int \left(\hat{p}(z')\widehat{A}\hat{p}^*(z)\vec{f},\vec{g}\right)_{H_0} d\mu(z') \\
&= \mu\left(\hat{p}(z')\widehat{A}\hat{p}^*(z)\vec{f},\vec{g}\right)_H = \mu\left(\vec{f},\hat{p}(z)\widehat{A}^*\hat{p}^*(z'\vec{g})\right)_H \\
&= \int \left(\vec{f},\hat{p}(z)\widehat{A}^*\hat{p}^*(z')\vec{g}\right)_{H_0} d\mu(z) = \left(\vec{f},\int \hat{p}(z)\widehat{A}^*\hat{p}^*(z')\widehat{P}\vec{g}d\mu(z)\right)_{H_0} \\
&= \left(\vec{f},\widehat{P}\widehat{A}^*\hat{p}^*(z)\widehat{P}\vec{g}\right)_{H_0}.
\end{aligned}$$

Hence, $\hat{v}^* = \widehat{P}\widehat{A}^*\hat{p}^*(z)\widehat{P}$.

LEMMA 3.5.3. *If one of the operators* $\widehat{T}_{1,2} = I + \widehat{A}_1\widehat{A}_2$, $\widehat{T}_{2,1} = I + \widehat{A}_2\widehat{A}_1$ *is invertible, then the other is also invertible, and,*

$$\widehat{T}_{\alpha,\beta}^{-1} = I - \widehat{A}_\alpha\widehat{T}_{\beta,\alpha}^{-1}\widehat{A}_\beta, \qquad \widehat{T}_{1,2}^{-1}\widehat{A}_1 = \widehat{A}_1\widehat{T}_{2,1}^{-1}.$$

PROOF: Assume that the operator $\widehat{T}_{1,2}$ is invertible. Then

$$\begin{aligned}
\widehat{T}_{2,1}(I - \widehat{A}_2\widehat{T}_{1,2}^{-1}\widehat{A}_1) &= I - \widehat{A}_2\widehat{T}_{1,2}^{-1}\widehat{A}_1 + \widehat{A}_2\widehat{A}_1 - \widehat{A}_2\widehat{A}_1\widehat{A}_2\widehat{T}_{1,2}^{-1}\widehat{A}_1 \\
&= I - \widehat{A}_2\widehat{T}_{1,2}^{-1}\widehat{A}_1 + \widehat{A}_2\widehat{A}_1 - \widehat{A}_2(\widehat{T}_{1,2} - I)\widehat{T}_{1,2}^{-1}\widehat{A}_1 = I, \\
(I - \widehat{A}_2\widehat{T}_{1,2}^{-1}\widehat{A}_1)\widehat{T}_{2,1} &= I - \widehat{A}_2\widehat{T}_{1,2}^{-1}\widehat{A}_1 + \widehat{A}_2\widehat{A}_1 - \widehat{A}_2\widehat{T}_{1,2}^{-1}\widehat{A}_1\widehat{A}_2\widehat{A}_1 \\
&= I - \widehat{A}_2\widehat{T}_{1,2}^{-1}\widehat{A}_1 + \widehat{A}_2\widehat{A}_1 - \widehat{A}_2\widehat{T}_{1,2}^{-1}(\widehat{T}_{1,2} - I)\widehat{A}_1 = I.
\end{aligned}$$

From this it follows that the operator $\widehat{T}_{2,1}$ is invertible and the equality $\widehat{T}_{2,1}^{-1} = I - \widehat{A}_2\widehat{T}_{1,2}^{-1}\widehat{A}_1$ is valid. The second equality follows from the identity

$$\widehat{T}_{1,2}\widehat{A}_1 = \widehat{A}_1 + \widehat{A}_1\widehat{A}_2\widehat{A}_1 = \widehat{A}_1(I + \widehat{A}_2\widehat{A}_1) = \widehat{A}_1\widehat{T}_{2,1}$$

and simultaneous from the invertibility of the operators $\widehat{T}_{1,2}, \widehat{T}_{2,1}$.

An operator $\widehat{I}$ defined in the entire space H_0 is said to be a conjugation operator if for all $\vec{f},\vec{g} \in H_0$,

$$(\widehat{I}\vec{f},\widehat{I}\vec{g}) = \overline{(\vec{f},\vec{g})}, \qquad \widehat{I}^2\vec{f} = \vec{f}.$$

A conjugation operator is antilinear $(\widehat{I}(\alpha\vec{f}_\beta\vec{g}) = \bar{\alpha}\widehat{I}(\vec{f}) + \bar{\beta}\widehat{I}(\vec{g}))$ and can be extended to the space $H = L^2_\mu(\Omega, H_0)$ by setting

$$\widehat{I}\left(\vec{f}(z)\right) = \sum \overline{f_k(z)}\widehat{I}(\vec{e}_k),$$

for all $\vec{f}(z) = \sum f_k(z)\vec{e}_k$. The extended operator will preserve its properties, i.e., it will remain a conjugation operator in the space H.

LEMMA 3.5.4. *Let the operator* $\widehat{T} = \widehat{C} + \widehat{R}\widehat{L}$ *be invertible,* $\widehat{L} = \widehat{L}^*$, *and* $\widehat{A} = \widehat{T}^{-1}\widehat{R}$.

1) If $\widehat{C} = I, \widehat{R} = \hat{a}\hat{r}\hat{b}$, *the operators* $\hat{a}, \hat{r}, \hat{b}$ *are self-adjoint, and* $[\hat{a}, \hat{b}] = [\hat{a}, \widehat{L}] = [\hat{b}, \widehat{L}] = 0$, *then the operator* $\hat{b}\widehat{A}\hat{a}$ *is self-adjoint:* $\hat{b}\widehat{A}\hat{a} = \hat{a}\widehat{A}^*\hat{b}$.

2) If a self-adjoint operator $\hat{a}$ *commutes with an operator* $\widehat{L}$ *and* $\widehat{R}\hat{a}\widehat{C}^* = \widehat{C}\hat{a}\widehat{R}^*$, *then the operator* $\widehat{A}\hat{a}$ *is self- adjoint:* $(\widehat{A}\hat{a})^* = \hat{a}\widehat{A}^* = \widehat{A}\hat{a}$.

3) If there exist such operators $\widehat{Q}, \widehat{D}$ *that* $[\widehat{D}, \widehat{L}] = 0$ *and* $\widehat{I}L\widehat{I} = \widehat{Q}\widehat{L}\widehat{Q}^{-1}$, $\widehat{I}\widehat{C}\widehat{I} = \widehat{Q}\widehat{D}\widehat{C}\widehat{D}^{-1}\widehat{Q}^{-1}$, $\widehat{I}\widehat{R}\widehat{I} = \widehat{Q}\widehat{D}\widehat{R}\widehat{D}^{-1}\widehat{Q}^{-1}$, *then* $\widehat{I}\widehat{A}\widehat{I} = \widehat{Q}\widehat{D}\widehat{A}\widehat{D}^{-1}\widehat{Q}^{-1}$.

PROOF: If $\widehat{C} = I, \widehat{R} = \hat{a}\hat{r}\hat{b}$, then, by Lemma 3.5.3, $\widehat{A} = \widehat{T}^{-1}\widehat{R} = \widehat{R}(I + \widehat{L}\widehat{R})^{-1}$, and since $\widehat{L} = \widehat{L}^*$, we have $\widehat{A}^* = (I + \widehat{R}^*\widehat{L})^{-1}\widehat{R}^*$, whence

$$\hat{a}\hat{r}\hat{b} = \widehat{R} = \widehat{T}\widehat{A} = \widehat{A} + \widehat{R}\widehat{L}\widehat{A} = \widehat{A} + \hat{a}\hat{r}\hat{b}\widehat{L}\widehat{A} = \widehat{A} + \hat{a}\hat{r}\widehat{L}\hat{b}\widehat{A},$$
$$\hat{b}\hat{r}\hat{a} = \widehat{R}^* = (I + \widehat{R}^*\widehat{L})\widehat{A}^* = \widehat{A}^* + \hat{b}\hat{r}\hat{a}\widehat{L}\widehat{A}^* = \widehat{A}^* + \hat{b}\hat{r}\widehat{L}\hat{a}\widehat{A}^*.$$

Multiply the former equality by $\hat{a}$ on the right and by $\hat{b}$ on the left, multiply the latter by $\hat{b}$ on the right and by $\hat{a}$ on the left, then subtract one from another. As a result, we arrive at the equality

$$\begin{aligned} 0 &= (\hat{b}\widehat{A}\hat{a} - \hat{a}\widehat{A}^*\hat{b}) + \hat{b}\hat{a}\hat{r}\widehat{L}(\hat{b}\widehat{A}\hat{a} - \hat{a}\widehat{A}^*\hat{b}) \\ &= (I + \hat{b}\hat{a}\hat{r}\widehat{L})(\hat{b}\widehat{A}\hat{a} - \hat{a}\widehat{A}^*\hat{b}). \end{aligned}$$

From Lemma 3.5.3 and invertibility of the operator $\widehat{T} = I + \hat{a}\hat{r}\hat{b}\widehat{L} = I + \hat{a}\hat{r}\widehat{L}\hat{b}$, it follows that the operator $I + \widehat{L}\hat{b}\hat{a}\hat{r}$ is invertible, whence, in turn, it follows that the operator $I + \hat{b}\hat{a}\hat{r}\widehat{L}$ is invertible, which implies the equality

$$\hat{b}\widehat{A}\hat{a} - \hat{a}\widehat{A}^*\hat{b} = 0.$$

If $\widehat{R}\hat{a}\widehat{C}^* = \widehat{C}\hat{a}\widehat{R}^*$ and $[\hat{a}, \widehat{L}] = 0$, then the equalities $\widehat{A} = \widehat{T}^{-1}\widehat{R}$, $\widehat{A}^* = R^*\widehat{T}^{*-1}$ imply that

$$\begin{aligned} \widehat{A}\hat{a} - \hat{a}\widehat{A}^* &= \widehat{T}^{-1}\widehat{R}\hat{a} - \hat{a}\widehat{R}^*\widehat{T}^{*-1} \\ &= \widehat{T}^{-1}(\widehat{R}\hat{a}\widehat{T}^* - \widehat{T}\hat{a}\widehat{R}^*)\widehat{T}^{*-1} \\ &= \widehat{T}^{-1}(\widehat{R}\hat{a}\widehat{C}^* - \widehat{R}\hat{a}\widehat{L}\widehat{R}^* - \widehat{C}\hat{a}\widehat{R}^* - \widehat{R}\widehat{L}\hat{a}\widehat{R}^*)\widehat{T}^{*-1} = 0, \end{aligned}$$

since all the terms within the parentheses cancel.

Finally, multiplying both parts of the equality $\widehat{R} = \widehat{C}\widehat{A} + \widehat{R}\widehat{L}\widehat{A}$ by the conjugation operator $\widehat{I}$ on the left and on the right, we find that

$$\begin{aligned} \widehat{I}\widehat{R}\widehat{I} &= \widehat{I}\widehat{C}\widehat{I}\widehat{I}\widehat{A}\widehat{I} + \widehat{I}\widehat{R}\widehat{I}\widehat{I}\widehat{L}\widehat{I}\widehat{I}\widehat{A}\widehat{I}, \\ \widehat{Q}\widehat{D}\widehat{R}\widehat{D}^{-1}\widehat{Q}^{-1} &= \widehat{Q}\widehat{D}\widehat{C}\widehat{D}^{-1}\widehat{Q}^{-1}\widehat{I}\widehat{A}\widehat{I} + \widehat{Q}\widehat{D}\widehat{R}\widehat{D}^{-1}\widehat{Q}^{-1}\widehat{Q}\widehat{L}\widehat{Q}^{-1}\widehat{I}\widehat{A}\widehat{I}, \\ \widehat{R} &= \widehat{C}(\widehat{D}^{-1}\widehat{Q}^{-1}\widehat{I}\widehat{A}\widehat{I}\widehat{Q}\widehat{D}) + \widehat{R}\widehat{L}(\widehat{D}^{-1}\widehat{Q}^{-1}\widehat{I}\widehat{A}\widehat{I}\widehat{Q}\widehat{D}). \end{aligned}$$

Therefore,

$$\widehat{D}^{-1}\widehat{Q}^{-1}\widehat{I}\widehat{A}\widehat{I}\widehat{Q}\widehat{D} = (\widehat{C} + \widehat{R}\widehat{L})^{-1}\widehat{R} = \widehat{A},$$

and hence, $\hat{I}\hat{A}\hat{I} = \hat{Q}\hat{D}\hat{A}\hat{D}^{-1}\hat{Q}^{-1}$.

Assume that the conditions of Lemma 3.5.1 are met. Then formula (3.5.1) can be rewritten in the following equivalent form:

$$\hat{v} = \hat{P}(\hat{\gamma} - iz)\hat{P} = -\hat{P}\hat{A}\hat{p}^*(z)\hat{P}, \qquad \hat{A} = \hat{T}^{-1}\hat{R}(z)\hat{\pi}_3, \qquad \hat{T} = I + \hat{R}(z)\hat{\pi}_3\hat{L}_1,$$

where $\hat{L}_1 = \hat{L}_1^*$. Whence, using Lemmas 3.5.2 and 3.5.4 (for the particular case when $\hat{C} = \hat{a} = \hat{b} = I$), we deduce that the operator $\hat{v} \in B(H_0)$ is self-adjoint if $(\hat{R}(z)\hat{\pi}_3)^* = \hat{R}(z)\hat{\pi}_3$. Since

$$(\hat{R}(z)\pi_3)^* = \hat{\pi}_3\hat{R}^*(z) = \hat{R}^*(-\bar{z})\hat{\pi}_3, \qquad \hat{R}(z) = \mathcal{E}(-2z)\hat{r}(z)$$

and $\mathcal{E}(2\bar{z})^* = \mathcal{E}(-2z)$, we see that the equality $(\hat{R}(z)\hat{\pi}_3)^* = \hat{R}(z)\hat{\pi}_3$ is satisfied if and only if $\hat{r}^*(-\bar{z}) = \hat{r}(z)$.

Thus, if the conditions of the Lemma are met and $\hat{r}(z) = \hat{r}^*(-\bar{z})$, we obtain self-adjoint solutions $\hat{u} = -\partial_x\hat{v} \in C^\infty(B(H_0))$ of the KdV equation.

Having carried out similar transformation in formula (3.5.2), we notice that the operators $a(z)$ and $\mathcal{E}_s^{-1}(z)\mathcal{E}_s(-z) = \exp\{a(-z)^s - a(z)^s\}t$ satisfy the condition $f(z) = f^*(-\bar{z})$; from this we conclude that the operators

$$\hat{v}_s = \hat{P}a(z)^{-s}\hat{\gamma}_s\hat{P} = I + \hat{P}a(z)^{-s}\hat{T}^{-1}\hat{R}(z)\hat{\pi}_3a(\bar{z})^s\hat{p}^*(z)\hat{P} \in B(H_0)$$

are self-adjoint if $\hat{r}_s(z) = \hat{r}_s^*(-\bar{z})$. Thus, under these conditions, we obtain self-adjoint solutions $\hat{v}_1 \in C^\infty(B(H_0))$ of the TL equation, while solutions $\hat{u}_2 = \hat{v}_2(n-1)^{-1}\hat{v}_2(n)$ of the LL equation will be real functions if the space H_0 is one-dimensional.

Let

$$\hat{r}(z) = \hat{a}\hat{r}_1(z)\hat{b}, \qquad \hat{r}_1(z) = \hat{r}_1^*(z)$$

in formula (3.5.3′) and let the self-adjoint operators $\hat{a}, \hat{b} \in B(H_0)$ commute with the operators $\hat{p}(z)$ and $\hat{B}$. Then, under the conditions of Lemma 3.5.1, formula (3.5.3) implies that

$$\hat{v} = \hat{P}\left(\hat{\gamma} - \hat{N}(z)\right)\hat{P} = -\hat{P}\hat{A}\hat{p}^*(z)\hat{P}\hat{B}, \qquad \hat{A} = \hat{T}^{-1}\hat{R}(z),$$

where

$$\hat{T} = I + \hat{R}(z)\hat{L}, \qquad \hat{L} = \hat{L}^*,$$
$$\hat{R}(z) = \hat{\mathcal{E}}(z)^{-1}\hat{a}\hat{r}_1(z)\hat{b}\hat{\mathcal{E}}^*(z)^{-1} = \hat{a}\hat{\mathcal{E}}(z)^{-1}\hat{r}_1(z)\hat{\mathcal{E}}^*(z)^{-1}\hat{b}.$$

Since the operators $\hat{T}, \hat{R}(z)$ meet condition 1 of Lemma 3.5.4, $(\hat{b}\hat{A}\hat{a})^* = \hat{b}\hat{A}\hat{a}$, and hence,

$$(\hat{b}\hat{v}\hat{a})^* = -(\hat{P}\hat{b}\hat{A}\hat{a}\hat{p}^*(z)\hat{P}\hat{B})^* = -\hat{B}\hat{P}(\hat{b}\hat{A}\hat{a})^*\hat{p}^*(z)\hat{P}$$
$$= -\hat{B}\hat{P}\hat{b}\hat{A}\hat{a}\hat{p}^*(z)\hat{P}\hat{B}\hat{B} = \hat{B}\hat{b}\hat{v}\hat{a}\hat{B} = \hat{b}\hat{B}\hat{v}\hat{B}\hat{a}.$$

Commuting both parts of the obtained equality $\hat{a}\hat{v}^*\hat{b} = (\hat{b}\hat{v}\hat{a})^* = \hat{b}\widehat{B}\hat{v}\widehat{B}\hat{a}$ with the operator $\widehat{B}$, we find that $a[\hat{v}^*, \widehat{B}]\hat{b} = \hat{b}\widehat{B}[\hat{v}, \widehat{B}]\widehat{B}\hat{a}$ and, since

$$[\hat{v}^*, \widehat{B}] = -[\hat{v}, B]^*, \widehat{B}[\hat{v}, \widehat{B}]\widehat{B} = -[\hat{v}, \widehat{B}],$$

we have $\hat{a}[\hat{v}, \widehat{B}]^*\hat{b} = \hat{b}[\hat{v}, \widehat{B}]\hat{a}$. So the operators

$$\hat{u} = [\hat{v}, \widehat{B}] = \widehat{P}[\hat{\gamma}, \widehat{B}]\widehat{P} \in C^\infty (B(H_0))$$

satisfy the relation $\hat{b}\hat{u}\hat{a} = \hat{a}\hat{u}^*\hat{b}$ and, depending on the choice of the function $\theta(z)$, are solutions to either the modified KdV equation or the nonlinear Schrödinger one.

Writing the operators in the operator matrix form, we see that under the assumptions made,

$$\hat{a} = \begin{pmatrix} \hat{a}_{11} & 0 \\ 0 & \hat{a}_{22} \end{pmatrix}, \quad \hat{b} = \begin{pmatrix} \hat{b}_{11} & 0 \\ 0 & \hat{b}_{22} \end{pmatrix}, \quad \hat{r}_1(z) = \begin{pmatrix} 0 & \hat{r}_{12}(z) \\ \hat{r}_{12}^*(z) & 0 \end{pmatrix},$$

$$\hat{r}(z) = \hat{a}\hat{r}_1(z)\hat{b} = \begin{pmatrix} 0 & \hat{a}_{11}\hat{r}_{12}(z)\hat{b}_{22} \\ \hat{a}_{22}\hat{r}_{12}^*(z)\hat{b}_{11} & 0 \end{pmatrix},$$

$$\hat{v} = \begin{pmatrix} \hat{v}_{11} & \hat{v}_{12} \\ \hat{v}_{21} & \hat{v}_{22} \end{pmatrix}, \qquad \hat{u} = [\hat{v}, \widehat{B}] = 2\begin{pmatrix} 0 & -\hat{v}_{12} \\ \hat{v}_{21} & 0 \end{pmatrix},$$

$$\hat{b}\hat{u}\hat{a} = 2\begin{pmatrix} 0 & -\hat{b}_{11}\hat{v}_{12}\hat{a}_{22} \\ \hat{b}_{22}\hat{v}_{21}\hat{a}_{11} & 0 \end{pmatrix}.$$

Here $\hat{b}_{22}\hat{v}_{21}\hat{a}_{11} = -\hat{a}_{22}\hat{v}_{12}^*\hat{b}_{11}$, as a consequence of the self-adjointness (proved above) of the operator $\hat{b}\hat{u}\hat{a}$. In particular, when $\hat{a}_{11} = I_1, \hat{b}_{22} = I_2$, we obtain the solutions of the form

$$\hat{u} = -2\begin{pmatrix} 0 & \hat{v}_{12} \\ \hat{a}_{22}\hat{v}_{12}^*\hat{b}_{11} & 0 \end{pmatrix},$$

if for $\hat{r}(z)$ we choose the operators of the same kind

$$\hat{r}(z) = \begin{pmatrix} 0 & \hat{r}_{12}(z) \\ \hat{a}_{22}\hat{r}_{12}^*(z)\hat{b}_{11} & 0 \end{pmatrix}.$$

Substituting the obtained expressions for the operators $\hat{u}$ into the corresponding nonlinear equations, we see that, in this case, the equations are reduced to

$$4\partial_t\hat{v}_{12} + \partial_x^3\hat{v}_{12} + 8\left\{\hat{v}_{12}\hat{a}_{22}\hat{v}_{12}^*\hat{b}_{11}(\partial_x\hat{v}_{12}) + (\partial_x\hat{v}_{12})\hat{a}_{22}\hat{v}_{12}^*\hat{b}_{11}\hat{v}_{12}\right\} = 0$$

if $\theta(z) = iz(x + z^2t)$, and

$$2i\partial_t\hat{v}_{12} + \partial_x^2\hat{v}_{12} + 8\hat{v}_{12}\hat{a}_{22}\hat{v}_{12}^*\hat{b}_{11}\hat{v}_{12} = 0,$$

if $\theta(z) = iz(x + zt)$.

Likewise it can be proved that under the conditions of Lemma 3.5.1 the operators

$$\hat{u} = \hat{P}(\hat{\gamma}_1 \hat{D}_1^{-1} - i\bar{z})\hat{P} = -\hat{P}\hat{T}_1^{-1}\hat{R}(z)\hat{p}^*(z)\hat{P}$$

satisfy the relation $\hat{a}\hat{u}^* = \hat{u}\hat{a}$ if $\hat{q}(z) = \hat{p}(z)^*$ in the formula (3.5.5) and if the operators $\hat{r}_0(z), \hat{Q}_1(z)$ are chosen so that

$$\hat{r}_0(z)\hat{a} = \hat{a}\hat{r}_0^*(z), \qquad \hat{Q}_1(z)\hat{a} = \hat{a}\hat{Q}_1^*(z),$$

where the self-adjoint operator $\hat{a} \in B(H_0)$ commutes with $\hat{p}(z), \hat{\mathcal{E}}(z)$. Therefore, for such a choice of parameters, we obtain the operators

$$\hat{u} = -\hat{P}\hat{T}_1^{-1}\hat{R}(z)\hat{p}^*(z)\hat{P} \in C^\infty\left(B(H_0)\right)$$

which solve the equation

$$\partial_x[\hat{u}, \hat{D}_2] - \partial_y[\hat{u}, \hat{D}_1] = \Big[[\hat{u}, \hat{D}_2], [\hat{u}, \hat{D}_1]\Big]$$

and satisfy the condition $\hat{a}\hat{u}^* = \hat{u}\hat{a}$ (if $\hat{a}\hat{r}_1^*(z) = \hat{r}_1(z)\hat{a}$ for real values of z).

Turning to formulas (3.5.6)and (3.5.6′), we find that under the conditions of Lemma 3.5.1,

$$\hat{v}_\alpha = \hat{P}\hat{\gamma}_\alpha A_\alpha^{-1}(\bar{z})\hat{D}_\alpha^{-1}\hat{P} = I + (-1)^\alpha \hat{P}\hat{T}_1^{-1}\hat{R}(z)A_\alpha(z)\hat{p}^*(z)\hat{P}$$

and, since $A_\alpha^*(z) = -A_\alpha(\bar{z})$, we have

$$\begin{aligned}\hat{v}_\alpha^* &= I + (-1)^\alpha \hat{P}A_\alpha^*(z)\left(\hat{T}_1^{-1}\hat{R}(z)\right)^*\hat{p}^*(z)\hat{P}\\ &= I - (-1)^\alpha \hat{P}A_\alpha(\bar{z})\hat{T}_1^{-1}\hat{R}(z)\hat{p}^*(z)\hat{P} = \hat{v}_\alpha^{-1},\end{aligned}$$

if the operator $\hat{T}_1^{-1}\hat{R}(z)$ is self-adjoint.

According to Lemma 3.5.4, operator $\hat{T}_1^{-1}\hat{R}(z)$ is self-adjoint if

$$\hat{q}(z) = \hat{p}^*(z), \qquad \hat{r}_0^*(z) = \hat{r}_0^*(z), \qquad \hat{Q}_1(z) = \hat{Q}_1^*(z).$$

Hence, if in formula (3.5.4′)

$$\hat{q}(z) = \hat{p}^*(z), \qquad \hat{r}_0(z) = \hat{r}_0^*(z), \qquad \hat{Q}_1(z) = \hat{Q}_1^*(z),$$

then the operators

$$\hat{u}_\alpha = \hat{v}_\alpha \hat{D}_\alpha \hat{v}_\alpha^{-1} = \hat{v}_\alpha \hat{D}_\alpha \hat{v}_\alpha^* \in C^\infty\left(B(H_0)\right)$$

are self-adjoint solutions of the system

$$\partial_y \hat{u}_1 = \frac{1}{2i}[\hat{u}_1, \hat{u}_2], \qquad \partial_x \hat{u}_2 = \frac{1}{2i}[\hat{u}_2, \hat{u}_1].$$

Further, the operator $\hat{v}$ defined by the equality (1.2.14′) equals $\widehat{T}_1^{-1} z \widehat{T}_1$ in the case considered, and since by the construction $z\widehat{T}_1 = \widehat{T}_1 \hat{z} + i\widehat{R}(z)\hat{p}^*(z)\widehat{P}$, then

$$\hat{z}^{-1}\hat{v} = I + i\hat{z}^{-1}\widehat{T}_1^{-1}\widehat{R}(z)\hat{p}^*(z)\widehat{P},$$
$$\hat{v}^{-1}\hat{z} = I - i\widehat{T}_1^{-1}z^{-1}\widehat{R}(z)\hat{p}^*\widehat{P}.$$

Hence,

$$\hat{w} = \widehat{P}\hat{z}^{-1}\hat{v}\widehat{P} = I + i\widehat{P}\hat{z}^{-1}\widehat{T}_1^{-1}\widehat{R}(z)\hat{p}^*(z)\widehat{P},$$
$$\hat{w}^{-1} = \widehat{P}\hat{v}^{-1}\hat{z}\widehat{P} = I - i\widehat{P}\widehat{T}_1^{-1}z^{-1}\widehat{R}(z)\hat{p}^*(z)\widehat{P};$$

from this, under the assumptions made above ($\hat{r}_0^*(z) = \hat{r}_0(z), \widehat{Q}_1(z) = \widehat{Q}_1^*(z)$), it follows that $\hat{w}^* = \hat{w}^{-1}$.

Thus, the operators $\hat{w} \in C^\infty(B(H_0))$ are unitary, while the operators

$$\widehat{S} = \hat{\mathcal{E}}(0)\hat{w} = \exp -i\left\{\hat{d}_1(x) + \hat{d}_2(y)\right\}\hat{w} \in C^\infty(B(H_0))$$

are unitary solutions of the equation

$$\partial_x\partial_y\widehat{S} = \frac{1}{2}\left\{(\partial_x\widehat{S})\widehat{S}^{-1}(\partial_y\widehat{S}) + (\partial_y\widehat{S})\widehat{S}^{-1}(\partial_x\widehat{S})\right\}.$$

Finally, formulas (3.5.3″) imply that

$$\hat{v} = \widehat{P}\widehat{N}(z)^{-1}\hat{\gamma}\widehat{P} = I - \widehat{P}N(z)^{-1}T^{-1}\widehat{R}(z)\widehat{B}\hat{p}^*(z)\widehat{P},$$
$$\hat{v}^{-1} = \widehat{P}\hat{\gamma}^{-1}\widehat{N}(z)\widehat{P} = I - \widehat{P}\widehat{B}\widehat{T}^{-1}\widehat{R}(z)\widehat{N}^*(z)^{-1}\hat{p}^*(z)\widehat{P},$$
$$\hat{v}^* = I - \widehat{P}\widehat{B}\left(\widehat{T}^{-1}\widehat{R}(z)\right)^*\widehat{N}^*(z)^{-1}\hat{p}^*(z)\widehat{P},$$

and, if in formulas (3.5.3′) the operator $\hat{r}(z)$ satisfies the condition

$$\hat{a}\hat{r}^*(z) = \hat{r}(z)\hat{a}, \qquad \left(\hat{a} = \hat{a}^* \in B(H_0), \qquad [\hat{a}, \hat{p}(z)] = [\hat{a}, \widehat{B}] = 0\right),$$

then, according to condition 2 of Lemma 3.5.4,

$$\begin{aligned}\hat{a}\hat{v}^* &= \hat{a} - \widehat{P}\widehat{B}\hat{a}\left(\widehat{T}^{-1}\widehat{R}(z)\right)^*\widehat{N}^*(z)^{-1}\hat{p}^*(z)\widehat{P}\\ &= \hat{a} - \widehat{P}\widehat{B}\widehat{T}^{-1}\widehat{R}(z)\hat{a}\widehat{N}^*(z)^{-1}\hat{p}^*(z)\widehat{P}\\ &= \left(I - \widehat{P}\widehat{B}\widehat{T}^{-1}\widehat{R}(z)\widehat{N}^*(z)^{-1}\hat{p}^*(z)\widehat{P}\right)\hat{a} = \hat{v}^{-1}\hat{a}.\end{aligned}$$

Thus, when $\theta(z) = i(zx - z^{-1}y)$, the operators $\hat{v} \in C^\infty(B(H_0))$ satisfy the relation $\hat{a}\hat{v}^* = \hat{v}^{-1}\hat{a}$ and the equation

$$\partial_y(\hat{v}^{-1}\partial_x\hat{v}) = \hat{v}^{-1}\widehat{B}\hat{v}\widehat{B} - \widehat{B}\hat{v}^{-1}\widehat{B}\hat{v}; \tag{3.5.9}$$

when $\theta(z) = i(zx + z^2t)$, the operators $\widehat{S} = \hat{v}\widehat{B}\hat{v}^{-1} \in C^\infty(B(H_0))$ satisfy the relations $\widehat{S}^2 = I$, $\hat{a}\widehat{S}^* = \widehat{S}\hat{a}$ and the equation

$$-4i\partial_t\widehat{S} = [\widehat{S}, \partial_x^2\widehat{S}].$$

Besides, if the measure is π_3-invariant $(d\mu(z) = d\mu(-\bar{z})), \theta(z) = i(zx - z^{-1}y)$, and the decomposable operators $\hat{m}(z), \hat{p}(z), \hat{r}(z)$ in formula (3.5.3′) satisfy the relations

$$\hat{I}\hat{m}(z)\hat{I} = \hat{m}(-\bar{z}), \qquad \hat{I}\hat{p}(z)\hat{I} = \hat{p}(-\bar{z}) = \hat{p}(z), \qquad \hat{I}\hat{r}(z)\hat{I} = \widehat{D}\hat{r}(-\bar{z})\widehat{D}^{-1},$$
$$(\widehat{D} \in B(H_0), \qquad [\widehat{D}, \hat{p}(z)] = [\widehat{D}, \widehat{B}] = 0),$$

then for $\widehat{Q} = \hat{\pi}_3$ condition 3 of Lemma 3.5.4 is met.

Therefore, in this case we obtain the solutions of Eq. (3.5.9) that satisfy the following two relations: $\hat{a}\hat{v}^* = \hat{v}^{-1}\hat{a}, \hat{I}\hat{v}\hat{I} = \widehat{D}\hat{v}\widehat{D}^{-1}$.

In conclusion we sum up the results obtained.

Let an operator $\widehat{T} = I + \widehat{R}(z)\widehat{L}$ be bounded and invertible.

1) If the operators $\widehat{R}(z), \widehat{L}$ are defined by formula (3.5.1′) and

$$d\mu(z) = d\mu(-\bar{z}), \qquad \chi_2(z)d\mu(z) = \chi_2(z)d\mu(-z),$$
$$\hat{q}(z) = \hat{p}^*(-\bar{z}), \qquad \hat{m}(z) = \hat{m}^*(\bar{z}), \qquad \hat{r}(z) = \hat{r}^*(-\bar{z}),$$

then the operators

$$\hat{u} = \hat{u}(x,t) = \partial_x\widehat{P}\widehat{T}^{-1}\widehat{R}(z)\hat{p}^*(-\bar{z})\widehat{P} \in C^\infty(B(H_0))$$

are self-adjoint and satisfy the equation

$$4\partial_t\hat{u} + 4\lambda\partial_x\hat{u} + \partial_x^3\hat{u} - 6\partial_x(\hat{u}^2) = 0.$$

2) If the operators $\widehat{R}(z), \widehat{L}$ defined by formula (3.5.2′) and

$$d\mu_s(z) = d\mu_s(-\bar{z}), \qquad \chi_2(z)d\mu_s(z) = \chi_2(z)d\mu_s(-z),$$
$$\hat{q}_s(z) = a(-z)\hat{p}_s^*(-\bar{z}), \qquad \hat{m}_s(z) = \hat{m}_s^*(\bar{z}), \qquad \hat{r}_s(z) = \hat{r}_s^*(-\bar{z}),$$

then the operators

$$\hat{v}_s = \hat{v}_s(t,n) = I + \widehat{P}a(-z)^s\widehat{T}^{-1}\widehat{R}(z)a(-z)^s\hat{p}^*(-\bar{z})\widehat{P} \in C^\infty(B(H_0))$$

are self-adjoint, the operators $\hat{v}_1$ satisfy the equation

$$\partial_t\left\{\hat{v}_1(n)^{-1}\partial_t v_1(n)\right\} = \hat{v}_1(n)^{-1}\hat{v}_1(n+1) - \hat{v}_1(n-1)\hat{v}_1(n),$$

and the operators $\hat{u} = \hat{u}(t,n) = \hat{v}_2(n-1)^{-1}\hat{v}_2(n)$ satisfy the equation

$$\partial_t\hat{u}(n) = \hat{u}(n)\hat{u}(n+1) - \hat{u}(n-1)\hat{u}(n),$$

provided that the following additional conditions are met:

$$d\mu_2(z) = d\mu_2(-z^{-1}), \qquad \hat{p}_2(z) = \hat{p}_2(-z^{-1}),$$
$$\hat{m}_2(z) = -\hat{m}_2(-z^{-1}), \qquad \hat{r}_2(z) = -\hat{r}_2(-z^{-1}).$$

3) Let the operators $\widehat{R}(z), \widehat{L}$ be defined by formula (3.5.3′) and

$$\bar{\chi}_2(z)d\mu(z) = \bar{\chi}_2(z)d\mu(\bar{z}), \qquad \hat{q}(z) = \hat{p}^*(z), \qquad \hat{m}(z) = \hat{m}^*(\bar{z}).$$

a) If in this formula

$$\hat{r}(z) = \begin{pmatrix} 0 & \hat{r}_{12}(z) \\ \hat{a}_{22}\hat{r}^*_{12}(z)\hat{b}_{11} & 0 \end{pmatrix},$$

and the operators

$$\hat{b}_{11} = \hat{b}^*_{11} \in B\left(\widehat{Q}_1(H_0)\right), \qquad \hat{a}_{22} = \hat{a}^*_{22} \in B\left(\widehat{Q}_2(H_0)\right)$$

commute with the operators $\widehat{Q}_1\hat{p}(z), \widehat{Q}_2\hat{p}(z)$, then the operator

$$u = \left[\widehat{P}\widehat{T}^{-1}\widehat{R}(z)\hat{p}^*(z)\widehat{P}\widehat{B}, \widehat{B}\right] \in C^\infty\left(B(H_0)\right)$$

has the form

$$\hat{u} = -2\begin{pmatrix} 0 & \hat{u}_{12} \\ \hat{a}_{22}\hat{u}^*_{12}\hat{b}_{11} & 0 \end{pmatrix},$$

while the operators $\hat{u}_{12} \in C^\infty\left(B(\widehat{Q}_2(H_0), \widehat{Q}_1(H_0))\right)$ satisfy the equation

$$4\partial_t\hat{u}_{12} + \partial_x^3\hat{u}_{12} + 8\left\{\hat{u}_{12}\hat{a}_{22}\hat{u}^*_{12}\hat{b}_{11}(\partial_x\hat{u}_{12}) + (\partial_x\hat{u}_{12})\hat{a}_{22}\hat{u}^*_{12}\hat{b}_{11}\hat{u}_{12}\right\} = 0$$

if $\theta(z) = iz(x + z^2t)$, or the equation

$$2i\partial_t\hat{u}_{12} + \partial_x^2\hat{u}_{12} + 8\hat{u}_{12}\hat{a}_{22}\hat{u}^*_{12}\hat{b}_{11}\hat{u}_{12} = 0,$$

if $\theta(z) = iz(x + zt)$.

b) If the operator $\hat{r}(z)$ in formula (3.5.3′) satisfies the relation

$$\hat{a}\hat{r}^*(z) = \hat{r}(z)\hat{a}(\hat{a} = \hat{a}^* \in B(H_0), [\hat{a}, \hat{p}(z)] = [\hat{a}, \widehat{B}] = 0),$$

then the operator

$$\hat{v} = I - \widehat{P}\widehat{N}(z)^{-1}\widehat{T}^{-1}\widehat{R}(z)\hat{p}^*(z)\widehat{P}\widehat{B} \in C^\infty\left(B(H_0)\right) \tag{3.5.10}$$

satisfies the relation $\hat{a}\hat{v}^* = \hat{v}^{-1}\hat{a}$ and the equation

$$\partial_y(\hat{v}^{-1}\partial_x\hat{v}) = \hat{v}^{-1}\widehat{B}\hat{v}\widehat{B} - \widehat{B}\hat{v}^{-1}\widehat{B}\hat{v}, \tag{3.5.10′}$$

if $\theta(z) = i(zx - z^{-1}y)$, while the operator $\widehat{S} = \hat{v}\widehat{B}\hat{v}^{-1}$ satisfies the relations $\widehat{S}^2 = I, \hat{a}\widehat{S}^* = \widehat{S}\hat{a}$ and the equation

$$4i\partial_t\widehat{S} = [\widehat{S}, \partial_x^2\widehat{S}],$$

if $\theta(z) = iz(x + zt)$.

c) If $\theta(z) = i(zx - z^{-1}y)$ in formula (3.5.3′) and

$$d\mu(z) = d\mu(-\bar{z}), \qquad \widehat{I}\hat{m}(z)\widehat{I} = \hat{m}(-\bar{z}), \qquad \widehat{I}\hat{p}(z)\widehat{I} = \hat{p}(-\bar{z}) = \hat{p}(z),$$
$$\widehat{I}\hat{r}(z)\widehat{I} = \widehat{D}\hat{r}(-\bar{z})\widehat{D}^{-1}, \qquad (\widehat{D} \in B(H_0), [\widehat{D}, \hat{p}(z)] = [\widehat{D}, \widehat{B}] = 0),$$

then the operator $\hat{v}$ defined by formula (3.5.10) satisfies Eq. (3.5.10′) and the relation $\widehat{I}\hat{v}\widehat{I} = \widehat{D}\hat{v}\widehat{D}^{-1}$.

Let the operator $\widehat{T}_1 = \widehat{C}(z) + \widehat{R}(z)\widehat{L}$ be bounded and invertible and the operators $\widehat{C}(z), \widehat{R}(z), \widehat{L}$ be defined by formulas (3.5.4′); also, let $\hat{q}(z) = \hat{p}^*(z)$.

4) If in formula (3.5.4′)

$$\mathcal{E}(z) = \exp iz\left(\hat{d}_1(x) + \hat{d}_2(y)\right), \qquad \hat{r}_0(z)\hat{a} = \hat{a}\hat{r}_0^*(z),$$
$$\widehat{Q}_1(z)\hat{a} = \hat{a}\widehat{Q}_1^*(z), \qquad \hat{a} = \hat{a}^* \in B(H_0),$$
$$[\hat{a}, \hat{d}_1(x)] = [\hat{a}, \hat{d}_2(y)] = [\hat{a}, \hat{p}(z)] = 0,$$

then the operator

$$\hat{u} = -\widehat{P}\widehat{T}_1^{-1}\widehat{R}(z)\hat{p}^*(z)\widehat{P} \in C^\infty\left(B(H_0)\right)$$

solves the equation

$$\partial_x[\hat{u}, \widehat{D}_2] - \partial_y[\hat{u}, \widehat{D}_1] = [[\hat{u}, \widehat{D}_2], [\hat{u}\widehat{D}_1]]$$

and satisfies the relation $\hat{a}\hat{u}^* = \hat{u}\hat{a}$.

5) If in formula (3.5.4′)

$$\mathcal{E}(z) = \exp\left\{A_1(z)\hat{d}_1(x) + A_2(z)\hat{d}_2(y)\right\},$$
$$A_1(z) = i(z-1)^{-1}, \qquad A_2(z) = -i(z+1)^{-1}$$
$$r_0(z) = \hat{r}_0^*(z), \qquad \widehat{Q}_1(z) = \widehat{Q}_1^*(z),$$

then the operators

$$\hat{v}_\alpha = I + (-1)^\alpha\widehat{T}_1^{-1}\widehat{R}(z)A_\alpha(z)\hat{p}^*(z)\widehat{P} \in C^\infty\left(B(H_0)\right) \qquad (\alpha = 1, 2)$$
$$\hat{w} = I + i\widehat{P}\bar{z}^{-1}\widehat{T}_1^{-1}\widehat{R}(z)\hat{p}^*(z)\widehat{P} \in C^\infty\left(B(H_0)\right),$$

are unitary, while the operators $\hat{u}_\alpha = \hat{v}_\alpha\widehat{D}_\alpha\hat{v}_\alpha^{-1} = \hat{v}_\alpha\widehat{D}_\alpha\hat{v}_\alpha^*$ are self-adjoint solutions of simultaneous equations

$$\partial_y\hat{u}_1 = \frac{1}{2i}[\hat{u}_1, \hat{u}_2], \qquad \partial_x\hat{u}_2 = \frac{1}{2i}[\hat{u}_2, \hat{u}_1],$$

while the operator $\widehat{S} = \exp -i\left\{\hat{d}_1(x) + \hat{d}_2(y)\right\}\hat{w}$ is a unitary solution of the equation

$$\partial_x\partial_y\widehat{S} = \frac{1}{2}\left\{(\partial_x\widehat{S})\widehat{S}^{-1}(\partial_y\widehat{S}) + (\partial_y\widehat{S})\widehat{S}^{-1}(\partial_x\widehat{S})\right\}.$$

§6 Invertibility Conditions for Operators $\widehat{\Gamma}$

In order to conclude the programme outlined in the beginning of the present chapter, we need to select invertible operators $\widehat{\Gamma}$ from the set of the operators $\widehat{\Gamma}$ found in the previous sections. Obviously, the operators $\widehat{\Gamma}$ are invertible and if and only if the operators $\widehat{T}$ or $\widehat{T}_1$ corresponding to them are invertible. Therefore it will suffice to find invertibility conditions for $\widehat{T}$ and T_1.

First, consider operators (3.5.1) related to the KdV equation. Employing results of the previous section, choose the parameters in this formula so that they would satisfy the conditions

$$\left.\begin{array}{ll} d\mu(z) = d\mu(-\bar{z}), & \chi_2(z)d\mu(z) = \chi_2(z)d\mu(-z), \\ \hat{r}(z) = \hat{r}^*(-\bar{z}), & \hat{m}(z) = \hat{m}^*(\bar{z}), \hat{q}(z) = \hat{p}^*(-\bar{z}) \end{array}\right\} \tag{3.6.1}$$

that guarantee self-adjointness of the solutions obtained to the KdV equation.

Boundedness criteria for the operators under consideration have been obtained in §4 of the present chapter.

Let us dwell briefly on the first of them in the case of one-dimensional space H_0. Let the measure $d\mu(z)$ be a restriction of the plane Lebesgue measure onto the compact Ω symmetric with respect to the imaginary axis. Setting $m(z) = 0$ and taking arbitrary continuous functions in the capacity of $p(z)$ and $r(z) = \overline{r(-\bar{z})}$, we obtain a bounded operator

$$\widehat{T} = I - \widehat{I}, \qquad \widehat{I}(f(z)) = -\int_\Omega \frac{\mathcal{E}(-z)^2 r(z)\overline{p(-\bar{z})}p(z')}{i(z'+z)} f(z')d\mu(z'),$$

that satisfies the reality condition (3.6.1). The kernel of the integral operator $\widehat{I}$ has a weak singularity, and the kernel of the operator $\widehat{I}^3$ is continuous and analytically depends on the variables x, t contained in the factor $\mathcal{E}(-z) = \exp -iz\left(x + (z^2 - \lambda)t\right)$. Therefore, the Fredholm determinant $\Delta(x,t) = \mathrm{Det}(I - \widehat{I}^3)$ exists and is an entire function of the variables x, t. If the norm of the function $r(z)$ is sufficiently small, then $\Delta(0,0) \neq 0$ and the determinant $\Delta(x,t)$ vanishes only on analytical curves. Outside these curves $\Delta(x,t) \neq 0$, and the operator $(I - \widehat{I}^3)$, together with it, the operator $I - \widehat{I}$ are invertible:

$$\left((I - \widehat{I})^{-1} = (I - \widehat{I}^3)^{-1}(I + \widehat{I} + \widehat{I}^2)\right).$$

Therefore the solutions of the KdV equation defined by formula (3.5.10) are real in the case considered, but they can have singularities (i.e., become infinite) on analytical curves of the (x,t) plane. Other possible variants in the case of the plane Lebesgue measure result in a similar situation. It seems impossible to get rid of singularities with such a choice of the measure.

Another criterion of boundedness of the operators $\widehat{\Gamma}$ was obtained for the case when the support Ω of the measure $d\mu(z)$ lies in the union of real and imaginary axes, its asymmetric part Ω_1 being contained in the finite number of intervals on the imaginary axis and satisfying condition b_3', while the measure $d\mu(z)$ on the set Ω_2 coincides with the linear Lebesgue one

$$(\chi_2(z)d\mu(z) = \chi_2(z)(2\pi^{-1}|dz|).$$

Note that in this case the constraints on the measure imposed by self-adjointness conditions (3.6.1) are met automatically for any choice of the measure on the set Ω_1.

Let us choose the operator functions $\hat{r}(z), \hat{m}(z), \hat{p}(z)$ and the Muckenhoupt weight $w(z)$, so that they would meet the remaining conditions of boundedness and self-adjointness. The latter imply that $q(z) = \hat{p}^*(-\bar{z}), \hat{r}_2(z) = \hat{r}(z)\hat{p}^*(-\bar{z})$, while it follows from conditions b_2, b_4 that the operator norms

$$\chi_1(z)\hat{r}_2(z) = \chi_1(z)\hat{r}(z)\hat{p}^*(-\bar{z}), \chi_1(z)\hat{p}(z), \chi_2(z)\hat{p}(z)\,(w(z))^{\frac{1}{2}},$$
$$\chi_2(z)\hat{r}_2(z)\,(w(-z))^{-\frac{1}{2}} = \chi_2(z)\hat{r}(z)\,(w(-\bar{z})w(-z))^{-\frac{1}{2}}\,\hat{p}^*(-\bar{z})\,(w(-\bar{z}))^{\frac{1}{2}}$$

must be bounded functions of z. These requirements are fulfilled if the Muckenhoupt weight meets the condition

$$\inf w(-\bar{z})w(-z) > 0,$$
$$\hat{p}(z) = \left(\chi_1(z) + \chi_2(z)w(z)^{-\frac{1}{2}}\right)\hat{p}_0(z),$$

and

$$\sup_z \|\hat{p}_0(z)\|_{H_0} < \infty, \qquad \sup_z \|\hat{r}(z)\|_{H_0} < \infty.$$

All boundedness conditions will be met if on the set Ω_1 one gives an arbitrary measure meeting condition b_5 with respect to the chosen Muckenhoupt weight. Note that if $\operatorname{dist}(\Omega_1, \Omega_2) > 0$, then condition b_5 is met for any measure.

We shall denote by $\chi_0(z), \chi_+(z), \chi_-(z)$ the characteristic functions of the sets formed by the intersection of the measure support Ω with the real axis $(\chi_0(z))$, the open upper half-plane $(\chi_+(z))$, and the open lower one $(\chi_-(z))$; we shall also denote the function $\varphi(z) = \chi_0(z) - \chi_+(z) + \chi_-(z)$ by $\varphi(z)$. Operators of multiplication by these functions will be denoted by $\chi_0, \chi_+, \chi_-, \varphi$. It is clear that $\chi_0 + \chi_+ + \chi_- = I$ and $\varphi^2 = I$.

To satisfy all the conditions of self-adjointness, we shall take for $\hat{m}(z)$ an arbitrary (for the time being) operator function that satisfies the conditions

$$\hat{m}(z) = \hat{m}^*(\bar{z}), \qquad \sup_z \|\hat{m}(z)\|_{H_0} < \infty,$$

while in the capacity of $\hat{r}(z)$, we shall take the operator function of the form

$$\hat{r}(z) = (\chi_0(z) - \chi_+(z) + \chi_-(z))\,\hat{r}_1(z)\hat{r}_1^*(-\bar{z}) = \varphi\hat{r}_1(z)\hat{r}_1^*(-\bar{z}),$$

where $\hat{r}_1(z)$ is an arbitrary invertible decomposable operator. At such a choice of the parameters

$$\widehat{R}(z) = \mathcal{E}(-z)^2\hat{r}(z) = \varphi\widehat{R}_1(z)\widehat{R}_1^*(-\bar{z})$$

with the invertible operator

$$\widehat{R}_1(z) = \mathcal{E}(-z)\hat{r}_1(z), \qquad \mathcal{E}(z) = \exp iz\left(x + (z^2 - \lambda)t\right).$$

Thus formula (3.5.1) can be rewritten as

$$\widehat{T} = \widehat{R}_1(z)\left[I + \varphi\widehat{R}_1^*(-\bar{z})\widehat{L}\widehat{R}_1(z)\right]\widehat{R}_1(z)^{-1},$$

whence it follows that the operator $\widehat{T}$ is invertible if and only if the operator

$$\widehat{A} = I + \varphi\widehat{R}_1^*(-\bar{z})\widehat{L}\widehat{R}_1(z) \tag{3.6.2}$$

is invertible.

Further considerations are based on the following simple Lemma.

LEMMA 3.6.1. *Let $\widehat{A} \in B(H)$, and there exists such a positive number d that* $\mathrm{Re}(\widehat{A}\vec{f}, \vec{f}) \geqslant d\|\vec{f}\|^2$ *for all $\vec{f} \in H$. Then the operator $\widehat{A}$ is invertible and* $\|\widehat{A}^{-1}\| \leqslant d^{-1}$.

PROOF: Having decomposed the operator $\widehat{A}$ into the sum of its real and imaginary components:

$$\widehat{A} = \widehat{A}_1 + i\widehat{A}_2, \qquad \widehat{A}_1 = \widehat{A}_1^* = \frac{1}{2}(\widehat{A} + \widehat{A}^*), \qquad \widehat{A}_2 = \widehat{A}_2^* = \frac{i}{2}(\widehat{A}^* - \widehat{A}),$$

we see that

$$\begin{aligned}(\widehat{A}_1\vec{f}, \vec{f}) &= \frac{1}{2}\left((\widehat{A} + \widehat{A}^*)\vec{f}, \vec{f}\right) = \frac{1}{2}\left\{(\widehat{A}\vec{f}, \vec{f}) + (\vec{f}, \widehat{A}\vec{f})\right\} \\ &= \mathrm{Re}(\widehat{A}\vec{f}, \vec{f}) \geqslant d\|\vec{f}\|^2.\end{aligned}$$

That equality shows that the real component of $\widehat{A}_1$ is a positive operator. So there exists a positive invertible operator $\widehat{C}$, such that $\widehat{A}_1 = \widehat{C}^2$, $\|\widehat{C}^{-1}\| \leqslant d^{-\frac{1}{2}}$, and

$$\widehat{A} = \widehat{C}^2 + i\widehat{A}_2 = \widehat{C}(I + i\widehat{C}^{-1}\widehat{A}_2\widehat{C}^{-1})\widehat{C}.$$

Since the operator $\widehat{C}^{-1}\widehat{A}_2\widehat{C}^{-1}$ is self-adjoint, the operator $I + i\widehat{C}^{-1}\widehat{A}_2\widehat{C}^{-1}$ is invertible and $\|(I + i\widehat{C}^{-1}\widehat{A}_2\widehat{C}^{-1})^{-1}\| \leqslant 1$. That is why the operator $\widehat{A}$ is also invertible:

$$\widehat{A}^{-1} = \widehat{C}^{-1}(I + i\widehat{C}^{-1}\widehat{A}_2\widehat{C}^{-1})^{-1}\widehat{C}^{-1},$$

and

$$\|\widehat{A}^{-1}\| = \|\widehat{C}^{-1}\|\|(I + i\widehat{C}^{-1}\widehat{A}_2\widehat{C}^{-1})^{-1}\|\|\widehat{C}^{-1}\| \leqslant d^{-1}.$$

Note that the conditions of the lemma impose constraints only on the real component of the operator; henceforth, we shall denote this component by $\operatorname{Re}\widehat{A} = \frac{1}{2}(\widehat{A} + \widehat{A}^*)$. Thus, testing the above conditions, one may consider, instead of the given operator, any operators that differ from it by anti-Hermitian terms $\widehat{D}(\widehat{D}^* = -\widehat{D})$.

Since

$$\widehat{R}_1^*(-\bar{z})\widehat{L}\widehat{R}_1(z) = \widehat{M}(z)\hat{\pi}_1 + \widehat{\Lambda},$$

where

$$\widehat{M}(z) = \chi_2(z)\widehat{R}_1^*(-\bar{z})\hat{m}(z)\widehat{R}_1(-z),$$

$$\widehat{\Lambda}\left(\vec{f}(z)\right) = \int \frac{\widehat{R}_1^*(-\bar{z})\hat{p}^*(-\bar{z})\hat{p}(z')\widehat{R}_1(z')}{i(z'+z)}\vec{f}(z')d\mu(z'),$$

and

$$\begin{aligned}\varphi\widehat{\Lambda} = (\chi_0 - \chi_+ + \chi_-)\widehat{\Lambda}(\chi_0 + \chi_+ + \chi_-) &= \chi_0\widehat{\Lambda}\chi_0 \\ &+ \{-\chi_+\widehat{\Lambda}\chi_+ + \chi_0\widehat{\Lambda}\chi_+ - \chi_+\widehat{\Lambda}\chi_0\} \\ &+ \chi_-\widehat{\Lambda}\chi_- + \chi_0\widehat{\Lambda}\chi_- + \chi_-\widehat{\Lambda}\chi_0\} \\ &+ \{\chi_-\widehat{\Lambda}\chi_+ - \chi_+\widehat{\Lambda}\chi_-\},\end{aligned}$$

where, as can be easily verified, the operator $\chi_-\widehat{\Lambda}\chi_+ - \chi_+\widehat{\Lambda}\chi_-$ is anti- Hermitian, we may take the operator

$$\begin{aligned}\widehat{A}_1 &= I + \varphi\widehat{M}(z)\pi_1 + \widehat{A}_0, \\ \widehat{A}_0 &= \chi_0\widehat{\Lambda}\chi_0 + \{-\chi_+\widehat{\Lambda}\chi_+ + \chi_0\widehat{\Lambda}\chi_+ - \chi_+\widehat{\Lambda}\chi_0\} \\ &\quad + \{\chi_-\widehat{\Lambda}\chi_- + \chi_0\widehat{\Lambda}\chi_- + \chi_-\widehat{\Lambda}\chi_0\}\end{aligned} \tag{3.6.3}$$

instead of operator (3.6.2).

Thus, to make sure that the operator (3.6.2) is invertible, it is sufficient to check whether the operator (3.6.3) meets the condition of Lemma 3.6.1.

Introduce the following notation:

$$\vec{a}_\pm(\lambda) = \int \chi_\pm(z)e^{i\lambda z}\hat{p}(z)\widehat{R}_1(z)\vec{f}(z)d\mu(z),$$

$$\vec{b}_\pm(\lambda) = \int \chi_0(z)e^{\pm i\lambda z}\hat{p}(\pm z)\widehat{R}_1(\pm z)\vec{f}(z)d\mu(z),$$

and note that

$$\vec{a}_\pm(\lambda) = \int \chi_\pm(z)e^{-i\lambda\bar{z}}\hat{p}(-\bar{z})\widehat{R}_1(-\bar{z})\vec{f}(z)d\mu(z),$$

because $z = -\bar{z}$ on the imaginary axis.

Combining this notation, the equality

$$\frac{1}{i(z'+z)} = \begin{cases} -\int_0^{\infty} e^{i\lambda(z'+z)}d\lambda, & \operatorname{Im}(z'+z) > 0 \\ \int_{-\infty}^{0} e^{i\lambda(z'+z)}d\lambda, & \operatorname{Im}(z'+z) < 0 \end{cases}$$

and the Plemelj-Sochocki formula

$$\begin{aligned} ⨍ \frac{\chi_0(y)g(y)}{i(y+x)}d\mu(y) &= \frac{1}{2\pi i} ⨍_{-\infty}^{\infty} \frac{g(y)\chi_0(y)}{y+x}dy \\ &= \frac{1}{2}\lim_{\varepsilon\to+0}\left\{\frac{1}{2\pi i}\int_{-\infty}^{\infty}\frac{g(y)\chi_0(y)}{y+x+i\varepsilon}dy + \frac{1}{2\pi i}\int_{-\infty}^{\infty}\frac{g(y)\chi_0(y)}{y+x-i\varepsilon}dy\right\} \\ &= -\frac{1}{4\pi}\lim_{\varepsilon\to+0}\int_{-\infty}^{\infty}\left\{\frac{\lambda}{|\lambda|}e^{i\lambda x-|\lambda|\varepsilon}\int_{-\infty}^{\infty}e^{i\lambda y}g(y)\chi_0(y)dy\right\}d\lambda \\ &= -\frac{1}{2}\int_{-\infty}^{\infty}\left\{\frac{\lambda}{|\lambda|}e^{i\lambda x}\int \chi_0(y)e^{i\lambda y}g(y)d\mu(y)\right\}d\lambda, \end{aligned}$$

we find that

$$\begin{aligned} \mu(\chi_0\widehat{\Lambda}\chi_0\vec{f}(z),\vec{f}(z))_H &= -\frac{1}{2}\int_{-\infty}^{\infty}\frac{\lambda}{|\lambda|}\left(\vec{b}_+(\lambda),\vec{b}_-(\lambda)\right)_{H_0}d\lambda, \\ \mu(\chi_\pm\widehat{\Lambda}\chi_\pm\vec{f}(z),\vec{f}(z))_H &= -\int_{-\infty}^{\infty}\chi_0^\pm(\lambda)\frac{\lambda}{|\lambda|}\left(\vec{a}_\pm(\lambda),\vec{a}_\pm(\lambda)\right)_{h_0}d\lambda, \\ \mu(\chi_\pm\widehat{\Lambda}\chi_0\vec{f}(z),\vec{f}(z))_H &= -\int_{-\infty}^{\infty}\chi_0^\pm(\lambda)\frac{\lambda}{|\lambda|}\left(\vec{b}_+(\lambda),\vec{a}_\pm(\lambda)\right)_{H_0}d\lambda, \\ \mu(\chi_0\widehat{\Lambda}\chi_\pm\vec{f}(z),\vec{f}(z)_H &= -\int_{-\infty}^{\infty}\chi_0^\pm(\lambda)\frac{\lambda}{|\lambda|}\left(\vec{a}_\pm(\lambda),\vec{b}_-(\lambda)\right)_{H_0}d\lambda, \end{aligned}$$

where by $\chi_0^+(\lambda)(\chi_0^-(\lambda))$ we denote the characteristic function of the positive (negative) part of the real axis. Calculate now $\operatorname{Re}(\widehat{A}_0\vec{f}(z),\vec{f}(z),\vec{f}(z))_H$. To do this, we shall use, in addition to the above equalities, the identity

$$\operatorname{Re}\left\{(a,a)+(a,b)+(c,a)+\frac{1}{2}(b,c)\right\} = \|a+\frac{b+c}{2}\|^2 - \frac{1}{4}\left(\|b\|^2+\|c\|^2\right),$$

which is valid for any scalar product.

Omitting intermediate calculations, we present the final result:

$$\begin{aligned} \mu\operatorname{Re}\left(\widehat{A}_0\vec{f}(z),\vec{f}(z)\right)_H &= -\frac{1}{4}\int_{-\infty}^{\infty}\left(\|\vec{b}_+(\lambda)\|_{H_0}^2+\|\vec{b}_-(\lambda)\|_{H_0}^2\right)d\lambda \\ &+\frac{1}{4}\int_{-\infty}^{\infty}\|\chi_0^+(\lambda)\left(2\vec{a}_+(\lambda)+\vec{b}_+(\lambda)-\vec{b}_-(\lambda)\right) \\ &+\chi_0^-(\lambda)\left(2\vec{a}_-(\lambda)+\vec{b}_+(\lambda)+\vec{b}_-(\lambda)\right)\|_{H_0}^2 d\lambda. \end{aligned}$$

From the Parseval equality for the Fourier transformations and from the definition of the functions $\vec{b}_\pm(\lambda)$, it follows that

$$\int_{-\infty}^{\infty} \|\vec{b}_\pm(\lambda)\|_{H_0}^2 \, d\lambda = \mu \|\chi_0(z)\hat{p}(z)\widehat{R}_1(z)\vec{f}(\pm z)\|_H^2 \leqslant \mu C_1 \|\vec{f}(z)\|_H^2,$$

and

$$\mathrm{Re}(\hat{A}_0\vec{f}(z), \vec{f}(z)) \geqslant -2C_1\|\vec{f}(z)\|_H^2,$$

where

$$C_1 = \sup_{-\infty<z<\infty} \|\hat{p}(z)\widehat{R}_1\|_{H_0}^2 = \sup_{-\infty<z<\infty} \|\hat{p}(z)\hat{r}_1(z)\|_{H_0}^2.$$

We have further

$$\begin{aligned}
&\mathrm{Re}(\varphi\widehat{M}(z)\hat{\pi}_1\vec{f}(z), \vec{f}(z))_H \\
&\quad = \mathrm{Re}\left((\chi_-(z) - \chi_+(z))\chi_2(z)\hat{r}_1^*(-\bar{z})\hat{m}(z)\hat{r}_1(-z)\hat{\pi}_1\vec{f}(z), \vec{f}(z)\right)_H \\
&\qquad + \mathrm{Re}\left(\chi_0(z)\hat{r}^*(-\bar{z})\hat{m}(z)\hat{r}(-z)\vec{f}(-z), \vec{f}(z)\right)_H \\
&\qquad\quad = \mathrm{Re}\left(\chi_0(z)\hat{r}^*(-\bar{z})\hat{m}(z)\hat{r}(-z)\vec{f}(-z), \vec{f}(z)\right)_H,
\end{aligned}$$

since under our conditions the operator

$$(\chi_-(z) - \chi_+(z))\,\chi_2(z)\hat{r}^*(-\bar{z})\hat{m}(z)\hat{r}, (-z)\hat{\pi}_1$$

is anti-Hermitian. Thus,

$$\mathrm{Re}\left(\varphi\widehat{M}(z)\hat{\pi}_1\vec{f}(z), \vec{f}(z)\right)_H \geqslant C_2\|\vec{f}(z)\|_H^2,$$

where

$$C_2 = \sup_{-\infty<z<\infty} \|\hat{r}^*(-\bar{z})\hat{m}(z)\hat{r}_1(-z)\|_{H_0}.$$

By comparing the results, we find

$$\mathrm{Re}\left(\widehat{A}\vec{f}(z), \vec{f}(z)\right)_H \geqslant \|\vec{f}(z)\|_H^2(1 - \frac{1}{2}C_1 - C_2).$$

So, if $\frac{1}{2}C_1 + C_2 < 1$, then the operator $\widehat{A}$ and hence $\widehat{T}$ are invertible with

$$\|\widehat{A}^{-1}\| \leqslant (1 - \frac{1}{2}C_1 - C_2)^{-1},$$

and this estimate of the norm of the operator $\widehat{A}^{-1}$ holds uniformly for all x, t.

Putting together these results, we arrive at the following theorem.

THEOREM 3.6.1. *Let*

1) the support Ω of the measure $d\mu(z)$ lie in the union of the real and imaginary axes, with its asymmetric part Ω_1 being contained in a finite number of intervals on the imaginary axis and satisfying condition b_3 (see §4);

2) the nonnegative function $w(z)$ satisfy the inequality $\inf_{z\in\Omega_2} w(z)w(\bar{z}) > 0$ and be a Muckenhoupt weight on the union of the real and imaginary axes;

3) the measure $d\mu(z)$ coincide with the linear Lebesgue measure ($d\mu(z) = (2\pi)^{-1}|dz|$) on the set Ω_2, while on the set Ω_1 it satisfies condition b_5 (see §4) with respect to the weight $w(z)$;

4) the operator functions $\hat{p}_0(z), \hat{r}_1(z), \hat{r}_1(z)^{-1}, \hat{m}(z) = \hat{m}^*(\bar{z})$ be bounded;

5) $\sup\limits_{-\infty<z<\infty} \|\hat{r}_1^*(-\bar{z})\hat{m}(z)\hat{r}_1(-z)\|_{H_0} + \frac{1}{2} \sup\limits_{-\infty<z<\infty} w(z)^{-1}\|\hat{p}_0(z)\hat{r}_1(z)\|_{H_0}^2 < 1.$

Then the operator $\widehat{T} = I + \widehat{R}(z)\widehat{L}$, where

$$\widehat{R}(z) = (\chi_0(z) - \chi_+(z) + \chi_-(z))\,\hat{r}_1(z)\hat{r}_1^*(-\bar{z}) \exp -2iz\left(x + (z^2 - \lambda)t\right),$$

$$\widehat{L}\left(\vec{f}(z)\right) = \chi_2(z)\hat{m}(z)\vec{f}(-z) + \int \frac{\hat{p}^*(-\bar{z})\hat{p}(z')}{i(z'+z)}\vec{f}(z')d\mu(z'),$$

$$\hat{p}(z) = \left(\chi_1(z) + \chi_2(z)w(z)^{-\frac{1}{2}}\right)\hat{p}_0(z)$$

is bounded and invertible, while the operators

$$\hat{u} = \hat{u}(x,t) = 2\partial_x \widehat{P}\widehat{T}^{-1}\widehat{R}(z)\hat{p}^*(-\bar{z})\widehat{P} \in C^\infty\left(B(H_0)\right)$$

are self-adjoint solutions of the KdV equation

$$4\partial_t\hat{u} + 4\lambda\partial_x\hat{u} + \partial_x^3\hat{u} - 3\partial_x(\hat{u}^2) = 0.$$

REMARK: If we replace the strict inequality in condition 5 by the equality, we shall have a more precise version of the theorem by imposing additional constraints that will allow us to employ the Fredholm theorems (see[25]).

LEMMA 3.6.2. *The operator $\widehat{T} = I + \widehat{A}\widehat{B}$ with $\widehat{A} = \widehat{A}^*, \widehat{B} = \widehat{B}^*$ is invertible if and only if the inequality*

$$\|\widehat{T}\vec{f}\| \geqslant d\|\vec{f}\|, \qquad d > 0, \tag{3.6.4}$$

holds for all vectors $\vec{f}$.

PROOF: Necessity of the condition is obvious. To prove sufficiency we note that the inequality (3.6.4) implies boundedness of the domain of values of the operator $\widehat{T}$, and since $(\widehat{T}\vec{x}, \vec{f}) = (\vec{x}, \widehat{T}^*\vec{f})$, we have that the orthogonal complement to this domain is the kernel of the conjugated operator $\widehat{T}^*$. If $\widehat{T}^*\vec{f} = \vec{f} + \widehat{B}\widehat{A}\vec{f} = 0$, then

$$\widehat{T}\widehat{A}\vec{f} = \widehat{A}\vec{f} + \widehat{A}\widehat{B}\widehat{A}\vec{f} = \widehat{A}(\vec{f} + \widehat{B}\widehat{A}\vec{f}) = 0,$$

and, according to (3.6.4), the vectors $\widehat{A}\vec{f}$ and $\vec{f} = -\widehat{B}\widehat{A}\vec{f}$ equal zero. Thus, the domain of values of the operator $\widehat{T}$ coincides with the entire space, and the operator is invertible.

Pass now to the operators $\widehat{T} = I + \widehat{R}(z)\widehat{L}$ defined by Eq. (3.5.3). We shall choose the parameters in this formula so that the boundedness conditions given in §3 and the self-adjointness conditions would be satisfied simultaneously. We shall use the notation:

C^+, C^- — open upper and lower half-planes, respectively;

R^1 — the real axis;

$\Omega^\pm = \Omega \cap C^\pm, \qquad \Omega^0 = \Omega \cap R^1, \qquad \overline{\Omega} = \pi_2(\Omega) = \{z : \bar{z} \in \Omega\}$;

$C^-(\infty), C^+(\infty)$ — connected components of complements to closures of the sets $\overline{\Omega}^+, \overline{\Omega}^-$ containing infinitely remote points;

$\Omega_2 = \Omega \cap \overline{\Omega} = \{z : z \in \Omega, \bar{z} \in \Omega\}, \Omega_1 = \Omega \backslash \Omega_2 = \{z : z \in \Omega, \bar{z} \notin \Omega\}, \Omega_i^\pm = \Omega_i \cap \Omega^\pm$,

$\chi_1(z), \chi_2(z)$ — characteristic functions of the sets Ω_1, Ω_2.

Note that characteristic functions of the sets $\Omega_i^\pm$ equal $\chi_i(z)\chi_\pm(z)$, while the characteristic function of the set Ω^0 is $\chi_0(z)$.

As a support of the measure $d\mu(z)$ we shall take the set Ω satisfying the following conditions:

1) The set Ω_2 consists of real axis segments and vertical straight lines.

2) The closure of the set Ω_1^+ lies in the connected component $C^+(\infty)$, while that of the set Ω_1^- lies in the connected component $C^-(\infty)$.

We define the measure $d\mu(z)$ on the set Ω_2 as a linear Lebesgue measure

$$\chi_2(z)d\mu(z) = \chi_2(z)(2\pi)^{-1}|dz|.$$

Obviously, it ensures that the first two conditions of boundedness stated in §3 are met and the equality $\chi_2(z)d\mu(z) = \chi_2(z)d\mu(\bar{z})$ is satisfied.

To ensure the remaining two conditions of boundedness and the condition of self-adjointness, we select the rest of the parameters in such a way that

3) $\chi_1(z)d\mu(z) \in K(w(x)) \cap K(w(x)^{-1})$

4)

$$\hat{p}(z) = \left(\chi_1(z) + \chi_2(z)w(z)^{-\frac{1}{2}}\right)\hat{p}_0(z), \qquad \hat{q}(z) = \hat{p}^*(z),$$

$$\hat{m}(z) = \hat{p}^*(z)\hat{\nu}(z)\hat{p}(z), \qquad \hat{\nu}(\bar{z}) = \hat{\nu}^*(z),$$

$$\hat{\nu}(z) = \chi_2(z)\hat{\nu}(z),$$

$$\hat{p}_0(z), \qquad \hat{\nu}(z), \hat{r}(z) \in L^\infty_\mu(\Omega, B(H_0)), \qquad \widehat{Q}_i\hat{p}_0(z) = \hat{p}_0(z)\widehat{Q}_i,$$

$$\widehat{Q}_i\hat{\nu}(z) = \hat{\nu}(z)\widehat{Q}_i, \qquad \hat{r}(z) = \widehat{Q}_1\hat{r}(z)\widehat{Q}_2 + \widehat{Q}_2\hat{r}(z)\widehat{Q}_1,$$

where $w(z)$ is the Muckenhoupt weight on the system of straight lines (mentioned in no. 1 above) meeting the condition

$$\inf_{z \in \Omega_2} w(z)w(\bar{z}) > 0.$$

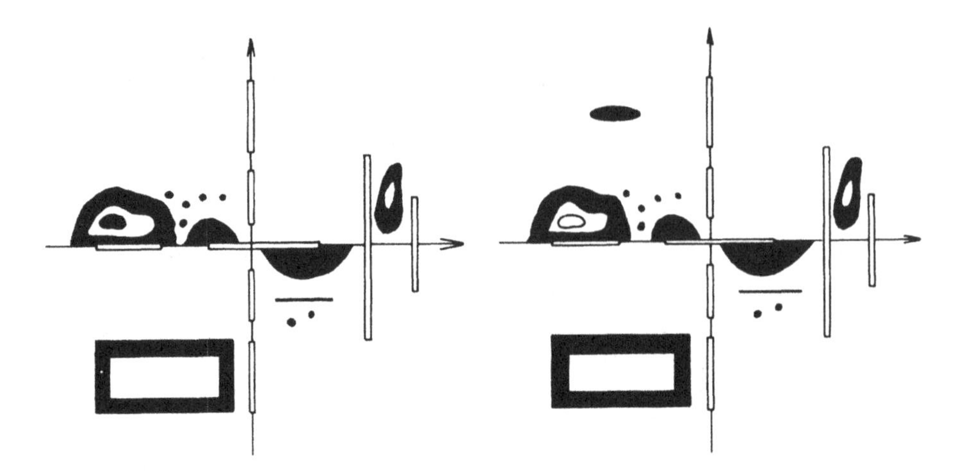

Figure 2

Figure 2 shows the set Ω meeting conditions 1 and 2 (left side) and meeting condition 1 but not 2 (right side). The notation is the same as Figure 1.

With such a choice of parameters, the operators $\widehat{L}$, $\widehat{R}(z)$ are bounded, $\widehat{L}^* = \widehat{L}$, and

$$\widehat{L} = \hat{p}^*(z)\widehat{L}_0\hat{p}(z), \qquad \widehat{L}_0 = \hat{\nu}(z)\hat{\pi}_2 + \widehat{\Lambda}, \qquad \widehat{\Lambda} = \widehat{\Lambda}_1\widehat{Q}_1 + \widehat{\Lambda}_2\widehat{Q}_2,$$

where

$$\widehat{\Lambda}_1\left(\vec{f}(z)\right) = \int \frac{\vec{f}(z')}{i(z'-\bar{z})}d\mu(z'), \qquad \widehat{\Lambda}_2\left(\vec{f}(z)\right) = \int \frac{\vec{f}(z')}{i(\bar{z}'-z)}d\mu(z')$$

are scalar integral operators.

First, we shall consider the case when the measure is concentrated on one of the vertical straights, so that $\Omega_1 = \emptyset$ and $\Omega = \Omega_2^+ \cup \Omega_2^-$. The identity $z' - \bar{z} = -(\bar{z}' - z)$, valid for any points z', z belonging to the same vertical line, implies that $\widehat{\Lambda}_2 = -\widehat{\Lambda}_1$ and, consequently, $\widehat{\Lambda} = \widehat{\Lambda}_1\widehat{B} = \widehat{B}\widehat{\Lambda}_1 (\widehat{B} = \widehat{Q}_1 - \widehat{Q}_2)$. Therefore,

$$\widehat{T} = I + \widehat{R}(z)\widehat{L} = I + \widehat{R}(z)\hat{p}_0^*(z)\widehat{B}\widehat{M_0}\hat{p}_0(z),$$

where

$$M_0 = \hat{\nu}_0(z)\hat{\pi}_2 + \widehat{\Lambda}_0, \qquad \hat{\nu}_0(z)\pi_2 = (w(z)w(\bar{z}))^{-\frac{1}{2}}\,\widehat{B}\hat{\nu}(z)\hat{\pi}_2,$$

and

$$\widehat{\Lambda}_0\left(\vec{f}(z)\right) = \int_\Omega \frac{(w(z)w(z'))^{-\frac{1}{2}}}{i(z'-\bar{z})}\vec{f}(z')d\mu(z')$$

are bounded self-adjoint operators. Then, if the operator $\hat{p}_0(z)$ is invertible, we have

$$\widehat{T} = \hat{p}_0(z)^{-1}\left\{I + \hat{\rho}(z)\widehat{B}\widehat{M_0}\right\}\hat{p}_0(z),$$

where

$$\hat{\rho}(z) = \hat{p}_0(z)\widehat{R}(z)\hat{p}_0(z)^* = \mathcal{E}(z)^{-1}\hat{p}_0(z)\hat{r}(z)\hat{p}_0^*(z)\mathcal{E}^*(z)^{-1}.$$

LEMMA 3.6.3. *Let the parameters in formula (3.5.3′) satisfy the above-stated conditions, let the measures $d\mu(z)$ be concentrated on one vertical line, and let $\Omega = \Omega_2^+ \cup \Omega_2^-$. Here are three conditions:*

1) the operator $\hat{\rho}_0(z)$ is invertible

$$(\hat{\rho}_0(z), \hat{\rho}_0(z)^{-1} \in L_\mu^\infty(\Omega, B(H_0)))\,;$$

2) $\hat{\rho}_0(z)\hat{\nu}(z) = -\hat{\nu}(z)\hat{\rho}_0(z), \qquad \hat{\rho}_0(\bar{z})\hat{\rho}_0(z) \geqslant 0;$

3) the operator

$$\widehat{D}\left(\vec{f}(z)\right) = \int\limits_\Omega w(z)^{-\frac{1}{2}} \frac{\hat{\rho}_0(z) - \hat{\rho}_0(\bar{z}')}{i(z' - \bar{z})} w(z')^{-\frac{1}{2}} \vec{f}(z')d\mu(z')$$

is compact.

If these conditions hold, then the homogeneous equation $\widehat{T}(\vec{f}(z)) = 0$ has a nonzero solution if and only if there exists such a sequence of unit vectors $\vec{f}(n,z)(\|\vec{f}(n,z)\|_H = 1)$ that

$$\lim_{n\to\infty} \|\widehat{T}\left(\vec{f}(n,z)\right)\|_H = 0.$$

PROOF: The necessity of the condition is obvious. If the sequence $\vec{f}(n,z)$ is compact, then, separating from it a convergent subsequence $\vec{f}(n,z)$ and setting $\vec{f}(z) = \lim_{n'\to\infty} \vec{f}(n',z)$, we find that $\|\vec{f}(z)\|_H = 1$ and

$$\widehat{T}(\vec{f}(z)) = \lim_{n'\to\infty} \widehat{T}(\vec{f}(n',z)) = 0.$$

Therefore, to prove sufficiency one must check that under the conditions of the lemma $\vec{f}(n,z)$ is compact. Since the operator $\hat{\rho}_0(z)$ is invertible, the sequence $\vec{f}(n,z)$ is compact if and only if the sequence

$$\vec{g}(n,z) = \hat{\rho}_0(z)\left(\vec{f}(n,z)\right)$$

is compact.

Since

$$(I + \hat{\rho}_0(z)\widehat{B}\widehat{M}_0)\,(\vec{g}(n,z)) = \hat{\rho}_0(z)\widehat{T}\left(\vec{f}(n,z)\right) \Rightarrow 0,$$

and

$$\widehat{B}\widehat{M}_0 = \widehat{M}_0\widehat{B}, \qquad \hat{\rho}_0(z)\widehat{B} = -\widehat{B}\hat{\rho}_0(z), \qquad \widehat{B}^2 = I,$$

then

$$\begin{aligned}
&(I - \hat{\rho}_0(z)\widehat{B}\widehat{M}_0)(I + \hat{\rho}_0(z)\widehat{B}\widehat{M}_0)\,(\vec{g}(n,z)) \\
&\quad = (I - \hat{\rho}_0(z)\widehat{B}\widehat{M}_0\hat{\rho}_0(z)\widehat{B}\widehat{M}_0)\,(\vec{g}(n,z)) \\
&\quad = (I + \hat{\rho}_0(z)\widehat{M}_0\hat{\rho}_0(z)\widehat{M}_0)\,(\vec{g}(n,z)) \Rightarrow 0.
\end{aligned}$$

Note that

$$\hat{\rho}_0(z)\widehat{M}_0 - \widehat{M}_0\hat{\rho}_0(z) = \hat{\rho}_0(z)\hat{\nu}_0(z)\hat{\pi}_2 - \hat{\nu}_0(z)\hat{\pi}_2\hat{\rho}_0(\bar{z}) + \hat{\rho}_0(z)\widehat{\Lambda}_0 - \widehat{\Lambda}_0\hat{\rho}_0(\bar{z})$$

and, according to condition 2,

$$\begin{aligned}\hat{\rho}_0(z)\hat{\nu}_0(z)\hat{\pi}_2 - \hat{\nu}_0(z)\hat{\pi}_2\hat{\rho}_0(\bar{z}) &= (w(z)w(\bar{z}))^{-\frac{1}{2}}\left\{\hat{\rho}_0(z)\widehat{B}\hat{\nu}(z) - \widehat{B}\hat{\nu}(z)\hat{\rho}_0(z)\right\}\hat{\pi}_2 \\ &= -(w(z)w(\bar{z}))^{-\frac{1}{2}}\,\widehat{B}\left\{\hat{\rho}_0(z)\hat{\nu}(z) + \hat{\nu}(z)\hat{\rho}_0(z)\right\}\hat{\pi}_2 = 0,\end{aligned}$$

while, according to condition 3,

$$\begin{aligned}&\left(\hat{\rho}_0(z)\widehat{\Lambda}_0 - \widehat{\Lambda}_0\hat{\rho}_0(\bar{z})\right)\left(\vec{f}(z)\right) \\ &\qquad = \int \frac{w(z)^{-\frac{1}{2}}\left(\hat{\rho}_0(z) - \hat{\rho}_0(\bar{z}')\right)w(z')^{-\frac{1}{2}}}{i(z' - \bar{z})}\vec{f}(z')d\mu(z') = \widehat{D}\left(\vec{f}(z)\right),\end{aligned}$$

where $\widehat{D}$ is a compact operator. Thus,

$$\hat{\rho}_0(z)\widehat{M}_0 = \widehat{M}_0\hat{\rho}_0(\bar{z}) + \widehat{D},$$
$$I + \hat{\rho}_0(z)\widehat{M}_0\hat{\rho}_0(z)\widehat{M}_0 = I + \widehat{M}_0\hat{\rho}_0(\bar{z})\hat{\rho}_0(z)\widehat{M}_0 + \widehat{D}\hat{\rho}_0(z)\widehat{M}_0$$

and, since $\widehat{M}_0 = \widehat{M}_0^*, \hat{\rho}_0(\bar{z})\hat{\rho}_0(z) \geqslant 0$ (condition 2), the operator

$$\widehat{M}_0\hat{\rho}_0(\bar{z})\hat{\rho}_0(z)\widehat{M}_0$$

is nonnegative. So the operator $\widehat{C} = I + \widehat{M}_0\hat{\rho}_0(\bar{z})\hat{\rho}_0(z)\widehat{M}_0$ is invertible, and

$$I + \hat{\rho}_0(z)\widehat{M}_0\hat{\rho}_0(z)\widehat{M}_0 = \widehat{C}(I + \widehat{C}^{-1}\widehat{D}\hat{\rho}_0(z)\widehat{M}_0) = \widehat{C}(I + \widehat{C}_1),$$

where $\widehat{C}_1 = \widehat{C}^{-1}\widehat{D}\hat{\rho}_0(z)\widehat{M}_0$ is a compact operator.

Finally, the relation

$$(I + \hat{\rho}_0(z)\widehat{M}_0\hat{\rho}_0(z)\widehat{M}_0)\left(\vec{g}(n,z)\right) \Rightarrow 0$$

implies that $(I + \widehat{C}_1)(\vec{g}(n,z)) \Rightarrow 0$. The fact that the sequences $\vec{g}(n,z), \vec{f}(n,z)$ are compact follows directly from the above relation and compactness of the operator $\widehat{C}_1$.

REMARK: Our assumption that the operator $\hat{\rho}_0(z)$ is invertible is inessential. Indeed, compactness of the sequence $\vec{f}(n,z)$ can be deduced from the equality

$$\widehat{T}\left(\vec{f}(n,z)\right) \equiv \vec{f}(n,z) + \widehat{R}(z)\hat{\rho}_0^*(z)\widehat{B}\widehat{M}_0\left(\vec{g}(n,z)\right) \Rightarrow 0$$

and compactness of the sequence $\vec{g}(n,z)$, which was shown to be compact without assuming invertibility of the operator $\hat{\rho}_0(z)$.

Passing to the general case, we shall first state some intermediate results. Since

$$\widehat{L}\varphi = (\chi_0 + \chi_+ + \chi_-)\widehat{L}(\chi_0 - \chi_+ + \chi_-) = \left\{\chi_+\widehat{L}(\chi_0 + \chi_-) - (\chi_0 + \chi_-)\widehat{L}\chi_+\right\} + \left\{\chi_0\widehat{L}\chi_0 - \chi_+\widehat{L}\chi_+ + \chi_-\widehat{L}\chi_- + \chi_0\widehat{L}\chi_- + \chi_-\widehat{L}\chi_0\right\},$$

and the operators $\widehat{L}, \chi_0, \chi_\pm$ are self-adjoint, we have

$$\operatorname{Re}\widehat{L}\varphi = \chi_0\widehat{L}\chi_0 - \chi_+\widehat{L}\chi_+ + \chi_-\widehat{L}\chi_- + \chi_0\widehat{L}\chi_- + \chi_-\widehat{L}\chi_0.$$

Besides, from the definition of the operator $\hat{\pi}_2$ and the identities

$$\chi_0(z)\chi_\pm(z) = \chi_0(z)\chi_\pm(\bar{z}) = \chi_\pm(z)\chi_\pm(\bar{z}) \equiv 0$$

it follows that

$$\operatorname{Re}\bar{\chi}_2(z)\hat{m}(z)\hat{\pi}_2\varphi = \chi_0(z)\hat{m}(z)\chi_0(z)$$

Introducing the notation

$$\vec{f}_i^{(0)}(\lambda) = \int \chi_0(z)e^{i\lambda z}\hat{p}_i(z)\vec{f}(z)d\mu(z),$$

$$\vec{f}_1^{\pm}(\lambda) = \int \chi_\pm(z)e^{i\lambda z}\hat{p}_1(z)\vec{f}(z)d\mu(z),$$

$$\vec{f}_2^{\pm}(\lambda) = \int \chi_\pm(z)e^{i\lambda z}\hat{p}_2(z)\vec{f}(z)d\mu(z),$$

and using the equalities obtained, we find, in the same way as in the proof of Theorem 3.6.1, that

(3.6.6)

$$\int \frac{\chi_\pm(z')\hat{p}_1(z')}{i(z'-\bar{z})}\vec{f}(z')d\mu(z') = -\int_0^{\pm\infty} \vec{f}_1^{\pm}(\lambda)e^{-i\lambda z}d\lambda, \qquad \operatorname{Im} \pm z \geqslant 0,$$

$$\int \frac{\chi_\pm(z')\hat{p}_2(z')}{i(\bar{z}'-z)}\vec{f}(z')d\mu(z') = \int_{\mp\infty}^0 \vec{f}_2^{\pm}(\lambda)e^{-i\lambda z}d\lambda, \qquad \operatorname{Im} \pm z \geqslant 0$$

$$-\mu\left(\chi_+\widehat{L}\chi_+\vec{f}_1(z), \vec{f}_1(z)\right)_H = \int_0^\infty \|\vec{f}_1^+(\lambda)\|_{H_0}^2 d\lambda,$$

$$\mu\left(\chi_+\widehat{L}\chi_+\vec{f}_2(z), \vec{f}_2(z)\right)_H = \int_{-\infty}^0 \|\vec{f}_2^+(\lambda)\|_{H_0}^2 d\lambda;$$

(3.6.7)

$$\begin{aligned}\operatorname{Re}&\left(\widehat{B}\widehat{L}\varphi\vec{f}(z), \vec{f}(z)\right)_H \\ &= \left(\chi_0(z)(\widehat{B}\hat{m}(z) - \frac{1}{2}\hat{p}^*(z)\hat{p}(z))\vec{f}(z), \vec{f}(z)\right)_H \\ &+ \mu^{-1}\left\{\int_0^\infty \left(\|\vec{f}_1^+(\lambda)\|_{H_0}^2 + \|\vec{f}_2^-(\lambda) + \vec{f}_2^0(\lambda)\|_{H_0}^2\right)d\lambda\right. \\ &+ \left.\int_{-\infty}^0 \left(\|\vec{f}_2^+(\lambda)\|_{H_0}^2 + \|\vec{f}_1^-(\lambda) + \vec{f}_1^0(\lambda)\|_{H_0}^2\right)d\lambda\right\}.\end{aligned}$$

where $\vec{f}_i(z) = \widehat{Q}_i(\vec{f}(z))$, $\widehat{B} = \widehat{Q}_1 - \widehat{Q}_2$.

THEOREM 3.6.2. *Let the parameters in formula (3.5.3′) satisfy the above-listed conditions.*

If, besides, $\hat{r}(z) = \hat{r}^*(z)$ *and*

1) $\chi_0(z)\left(\widehat{B}\hat{m}(z) - \frac{1}{2}\hat{p}^*(z)\hat{p}(z)\right) \geqslant 0$

2) restrictions of the operator functions $\hat{r}(z), \hat{p}(z), \hat{\nu}(z)$ *on the set* $\Omega_2^+ \cup \Omega_2^-$ *satisfy the conditions of Lemma 3.6.3, then the operator* $\widehat{T} = I + \widehat{R}(z)\widehat{L}$ *is bounded and invertible.*

PROOF: Self-adjointness of the operator $\hat{r}(z)$ implies that the operator $\widehat{R}(z) = \widehat{\mathcal{E}}(z)^{-1}\hat{r}(z)\widehat{\mathcal{E}}^*(z)^{-1}$ is self-adjoint too. Hence, according to Lemma 3.6.2, the operator $\widehat{T}$ is invertible if it satisfies the inequality

$$\inf_{\|\vec{f}\|=1} \|\widehat{T}\left(\vec{f}(z)\right)\|_H = d > 0.$$

Validity of the inequality is proved by contradiction. If $d = 0$, then there exists a sequence of vectors $\vec{f}(n,z)$ such that $\|\vec{f}(n,z)\|_H = 1$ and

$$\lim_{n\to\infty} \|\widehat{T}\left(\vec{f}(n,z)\right)\|_H = 0. \tag{3.6.8}$$

Since $\widehat{B}\widehat{L}\varphi\widehat{T} = \widehat{B}\widehat{L}\varphi + \widehat{L}\varphi\widehat{B}\widehat{R}(z)\widehat{L}$ and the operator $\varphi\widehat{B}\widehat{R}(z) = \varphi(\widehat{Q}_1\widehat{R}(z)\widehat{Q}_2 - \widehat{Q}_2\widehat{R}(z)\widehat{Q}_1)$ is anti-Hermitian, then $\operatorname{Re}\widehat{B}\widehat{L}\varphi\widehat{T} = \operatorname{Re}\widehat{B}\widehat{L}\varphi$ and, therefore,

$$\lim_{n\to\infty}\left(\operatorname{Re}\widehat{B}\widehat{L}\varphi\widehat{T}\vec{f}(n,z), \vec{f}(n,z)\right)_H = \lim_{n\to\infty}\operatorname{Re}\left(\widehat{B}\widehat{L}\varphi\vec{f}(n,z), \vec{f}(n,z)\right)_H = 0. \tag{3.6.9}$$

Equality (3.6.7) shows that the operator $\operatorname{Re}\widehat{B}\widehat{L}\varphi$ is nonnegative if condition 1 is met, and, since for any nonnegative operators $\widehat{A}$ the inequality $\|\widehat{A}\vec{f}\|^2 \leqslant \|\widehat{A}\|(\widehat{A}\vec{f}, \vec{f})$ is true, we have

$$\lim_{n\to\infty} \|\operatorname{Re}\widehat{B}\widehat{L}\varphi\vec{f}(n,z)\|_H = 0 \tag{3.6.10}$$

Recalling the definition of the function $\varphi(z)$, we find that $\widehat{L} = \operatorname{Re}\widehat{L}\varphi + \chi_+\widehat{L} + \widehat{L}\chi_+$, and, since $\widehat{B}^2 = I$, we have

$$\widehat{T} = I + \widehat{R}(z)\widehat{L} = I + \widehat{R}(z)(\chi_+\widehat{L} + \widehat{L}\chi_+) + \widehat{R}(z)\widehat{B}\operatorname{Re}\widehat{B}\widehat{L}\varphi,$$

whence, according to Eq. (3.6.10), we deduce that

$$\lim_{n\to\infty}\|\widehat{T}\vec{f}(n,z)\|_H = \lim_{n\to\infty}\|\vec{f}(n,z) + \widehat{R}(z)(\chi_+\widehat{L} + \widehat{L}\chi_+)\vec{f}(n,z)\|_H = 0. \tag{3.6.11}$$

Further, equality (3.6.9) and condition 1 imply that

$$\lim_{n\to\infty}\left\{\int_0^\infty \|\vec{f}_1^+(n,\lambda)\|^2_{H_0}d\lambda + \int_{-\infty}^0 \|\vec{f}_2^+(n,\lambda)\|^2_{H_0}d\lambda\right\} = 0, \tag{3.6.12}$$

whence, employing formula (3.6.6), we arrive at the following equalities

$$\lim_{n\to\infty} \chi_+ \widehat{L} \chi_+ \vec{f}(n,z) = \lim_{n\to\infty} \chi_0 \widehat{L} \chi_+ \vec{f}(n,z) = 0.$$

Since

$$\chi_+ \widehat{L} + \widehat{L} \chi_+ = 2\chi_+ \widehat{L} \chi_+ + \chi_0 \widehat{L} \chi_+ + \chi_+ \widehat{L} \chi_0 + \chi_+ \widehat{L} \chi_- + \chi_- \widehat{L} \chi_+,$$

the equalities obtained permit us to replace formula (3.6.11) by the following one:

$$\lim_{n\to\infty} \left\{ \vec{f}(n,z) + \widehat{R}(z)(\chi_+ \widehat{L} \chi_0 + \chi_+ \widehat{L} \chi_- + \chi_- \widehat{L} \chi_+) \vec{f}(n,z) \right\} = 0,$$

whence, using the identity

$$\chi_0 \widehat{R}(z)(\chi_+ \widehat{L} \chi_0 + \chi_+ \widehat{L} \chi_- + \chi_- \widehat{L} \chi_+) = 0,$$

we conclude that

$$\begin{aligned} &\lim_{n\to\infty} \chi_0(z) \vec{f}(n,z) = 0, \\ &\lim_{n\to\infty} \left\{ \vec{f}(n,z) + \widehat{R}(z)(\chi_+ \widehat{L} \chi_- + \chi_- \widehat{L} \chi_+) \vec{f}(n,z) \right\} = 0, \end{aligned} \tag{3.6.13}$$

and

$$\lim_{n\to\infty} \left\{ \chi_\mp(z) \vec{f}(n,z) + \widehat{R}(z) \chi_\mp \widehat{L} \chi_\pm \vec{f}(n,z) \right\} = 0, \tag{3.6.14}$$

where

$$\begin{aligned} \chi_\mp \widehat{L} \chi_\pm \vec{f}(n,z) = {} & \chi_2(z) \chi_\mp(z) \chi_\pm(z) \widehat{m}(z) \vec{f}(n,z) \\ & + \chi_\mp(z) \widehat{p}^*(z) \left\{ \int \frac{\chi_\pm(z') \widehat{p}_1(z')}{i(z' - \bar{z})} \vec{f}(n,z') d\mu(z') \right. \\ & \left. + \int \frac{\chi_\pm(z') \widehat{p}_2(z')}{i(\bar{z}' - z)} \vec{f}(n,z') d\mu(z') \right\}. \end{aligned} \tag{3.6.14'}$$

The vector functions

$$\begin{aligned} \vec{g}_1(n,\bar{z}) &= \int \frac{\chi_+(z') \widehat{p}_1(z')}{i(z' - \bar{z})} \vec{f}(n,z') d\mu(z'), \\ \vec{g}_2(n,z) &= \int \frac{\chi_+(z') \widehat{p}_2(z')}{i(\bar{z}' - z)} \vec{f}(n,z') d\mu(z') \end{aligned}$$

are holomorphic in the domain $C^-(\infty)$, uniformly bounded on its every compact set and, according to Eqs. (3.6.6) and (3.6.12), tend to zero when $n \to \infty$ in the upper half-plane. Thus, when $n \to \infty$, they uniformly tend to zero on each compact of the domain $C^-(\infty)$, in particular, on the set $\Omega_1^- \subset C^-(\infty)$. This implies, according to Eqs. (3.6.14) and (3.6.14′), that

$$\lim_{n\to\infty} \bar{\chi}_1(z)\chi_-(z)\vec{f}(n,z) = 0$$

since $\bar{\chi}_1(z)\bar{\chi}_2(z) \equiv 0$.

The equality

$$\lim_{n\to\infty} \bar{\chi}_1(z)\chi_+(z)\vec{f}(n,z) = 0$$

is proved in the same manner. Comparing the equalities obtained with Eq. (3.6.13), we find that

$$\vec{f}(n,z) = \bar{\chi}_2(z)\left(\chi_+(z) + \chi_-(z)\right)\vec{f}(n,z) + \vec{\varepsilon}(n,z), \tag{3.6.15}$$

where

$$\vec{\varepsilon}(n,z) = \chi_0(z)\vec{f}(n,z) + \bar{\chi}_1(z)\left(\chi_+(z) + \chi_-(z)\right)\vec{f}(n,z)$$

and

$$\lim_{n\to\infty} \|\vec{\varepsilon}(n,z)\|_H = 0.$$

By the condition, the set $\Omega_2^+ \cup \Omega_2^-$ lies on the finite system of vertical lines. Denote the parts of this set lying on the kth straight line by $\Omega_2^+(k), \Omega_2^-(k)$; denote the loop surrounding the set Ω_2^- and containing no other points of the set $\overline{\Omega}^+$ by $\mathcal{L}_k$. Consider the integral

$$\begin{aligned}
\frac{1}{2\pi i}\int_{\mathcal{L}_k} \frac{\vec{g}_2(n,\xi)}{\xi - w}\,d\xi &= \frac{1}{2\pi i}\int_{\mathcal{L}_k} \frac{1}{\xi - w}\left(\int \frac{\chi_+(z')\hat{p}_2(z')}{i(\bar{z}' - \xi)}\vec{f}(n,z')\,d\mu(z')\right) d\xi \\
&= \frac{1}{2\pi i}\int_{\mathcal{L}_k} \frac{1}{\xi - w}\left\{\int_{\Omega_2^+(k)} \frac{\hat{p}_2(z')\vec{f}(n,z')}{i(\bar{z}' - \xi)}\,d\mu(z')\right. \\
&\qquad \left. + \int_{\Omega^+\setminus\Omega_2^+(k)} \frac{\hat{p}_z(z')\vec{f}(n,z')}{i(\bar{z}' - \xi)}\,d\mu(z')\right\} d\xi,
\end{aligned}$$

where w is an arbitrary point of the upper half-plane lying outside the loop $\mathcal{L}_k$. Since the second integral in the braces is a holomorphic vector function of ξ inside the loop $\mathcal{L}_k$, the second term of the right-hand side of this equality is equal to zero. Applying the Cauchy integral formula to the first term of the right-hand side, we find that

$$\frac{1}{2\pi i}\int_{\mathcal{L}_k} \frac{\vec{g}_2(n,\xi)}{\xi - w}\,d\xi = \int_{\Omega_2^+(k)} \frac{\hat{p}_2(z')\vec{f}(n,z')}{i(\bar{z}' - w)}\,d\mu(z')$$

and, since the vector functions $\vec{g}_2(n,\xi)$ tend to zero on the loop $\mathcal{L}_k$ uniformly, then

$$\lim_{n\to\infty} \int_{\Omega_2^+(k)} \frac{\hat{p}_2(z')\vec{f}(n,z')}{i(\bar{z}' - w)}\,d\mu(z') = 0. \tag{3.6.16}$$

The equalities

$$\lim_{n\to\infty}\int\limits_{\Omega_2^+(k)}\frac{\hat{p}_1(z')\vec{f}(n,z')}{i(z'-\bar{w})}\,d\mu(z')$$

$$(3.5.16')\qquad =\lim_{n\to\infty}\int\limits_{\Omega_2^-(k)}\frac{\hat{p}_2(z')\vec{f}(n,z')}{i(\bar{z}'-w)}\,d\mu(z')$$

$$=\lim_{n\to\infty}\int\limits_{\Omega_2^-(k)}\frac{\hat{p}_1(z')\vec{f}(n,z')}{i(z'-\bar{w})}\,d\mu(z')=0$$

are proved in a similar fashion.

Denote the characteristic functions of the sets $\Omega_2^+(k)\cup\Omega_2^-(k)$ by $\chi_k(z)$, and by χ_k we denote the operators of multiplication by these functions. Since $\sum_k \chi_k(z) = \bar{\chi}_2(z)\,(\chi_+(z)+\chi_-(z))$, we obtain, by combining Eqs. (3.6.8) and (3.6.15), that

$$\lim_{n\to\infty}\left\{\chi_l\vec{f}(n,z)+\widehat{R}(z)\chi_l L\Big(\sum_k \chi_k\Big)\vec{f}(n,z)\right\}=0.$$

On the other hand, Eqs. (3.6.16) and (3.6.16′) imply that

$$\lim_{n\to\infty}\chi_l\widehat{L}\chi_k\vec{f}(n,z)=0,$$

if $k\neq l$. Therefore,

$$\lim_{n\to\infty}\left\{\chi_l\vec{f}(n,z)+\widehat{R}(z)\chi_l\widehat{L}\chi_l\vec{f}(n,z)\right\}=0,$$

and, by Lemma 3.6.3, the sequences $\chi_l\vec{f}(n,z)$ are compact. Thus, it is possible to separate from the sequence $\vec{f}(n,z)=\sum_l\chi_l\vec{f}(n,z)$ a subsequence that converges to the vector $\vec{f}(z)$, while $\widehat{T}(\vec{f}(z))=0$ and $\|\vec{f}(z)\|_H=1$. Passing to the limit in the formulas defining the vector functions $\vec{g}_1(n,z),\vec{g}_2(n,z)$, the limit being taken over the above subsequence, we arrive at the equalities

$$\int\frac{\chi_+(z')\hat{p}_1(z')}{i(z'-\bar{z})}\vec{f}(z')d\mu(z')=\int\frac{\chi_+(z')\hat{p}_2(z')}{i(\bar{z}'-z)}\vec{f}(z')d\mu(z')\equiv 0,$$

whence, according to Plemelj–Sochocki formulas, we get

$$\chi_+(z')\left\{\hat{p}_1(z')+\hat{p}_2(z')\right\}\left(\vec{f}(z')\right)\equiv 0.$$

Finally, this identity and Eq. (3.6.14) imply that $\chi_\mp(z)\vec{f}(z)$, hence, $\vec{f}(z)$ equal zero.

Thus, after all, the assumption that a sequence satisfying condition (3.6.8) exists leads to a contradiction: $\|\vec{f}(z)\|_H = 1$ and $\vec{f}(z) \equiv 0$.

Now, let $\hat{r}(z) \neq \hat{r}^*(z)$. Sufficient condition of invertibility of the operator $\widehat{T}$ is easy to obtain in the case when the measure $d\mu(z)$ is concentrated on the real axis and $d\mu(z) = (2\pi)^{-1}|dz|$. Indeed, for real values of z

$$\|\widehat{R}(z)|_{H_0} = \|\hat{r}(z)\|_{H_0},$$
$$\|\widehat{R}(z)\hat{m}(z)\|_{H_0} \leqslant \|\hat{r}(z)\|_{H_0}\|\hat{m}(z)\|_{H_0},$$
$$\|\widehat{R}(z)\hat{p}^*(z)\|_{H_0} \leqslant \|\hat{r}(z)\|_{H_0}\|\hat{p}(z)\|_{H_0}.$$

Therefore,

$$\|\widehat{R}(z)\hat{m}(z)\|_H \leqslant C(r,m) = \sup_{-\infty<z<\infty} \|\hat{r}(z)\|_{H_0}\|\hat{m}(z)\|_{H_0},$$
$$\|\widehat{R}(z)\hat{p}^*(z)\|_H \leqslant C(r,p) = \sup_{-\infty<z<\infty} \|\hat{r}(z)\|_{H_0}\|\hat{p}(z)\|_{H_0},$$

whence

$$\|\widehat{R}(z)\widehat{L}\|_H \leqslant C(r,m) + C(r,p)D,$$

whereby D we denote the norm of the operator

$$\widehat{D}\left(\vec{f}(z)\right) = \int \frac{\hat{p}(z')\vec{f}(z')}{i(z'-z)}\,d\mu(z') \qquad (-\infty < z < \infty).$$

Passing to the Fourier transformation, we find that

$$\|\widehat{D}\|_H = D \leqslant \sup_{-\infty<z<\infty} \|\hat{p}(z)\|_{H_0} = C(p).$$

Thus,

$$\|\widehat{R}(z)\widehat{L}\|_H \leqslant C(r,m) + C(r,p)C(p)$$

and, if

$$C(r,m) + C(r,p)C(p) < 1$$

then the operator $\widehat{T}$ is invertible.

CHAPTER 4

Classes Of Solutions To Nonlinear Equations

The results obtained in the previous chapter enable us to classify the nonlinear equations under consideration. The properties of these solutions depend on the choice of parameters contained in the operators $\hat{\Gamma}$. It is this dependence that is investigated in the present chapter. For the sake of simplicity, the space H_0 is assumed to be of the finite dimension n. We shall choose an orthonormal basis $\vec{e_1}, \vec{e_2}, \ldots, \vec{e_n}$ and shall denote the operator matrices $\hat{A} \in B(H_0)$ and operator functions $\hat{A}(\omega)$ in this basis simply by A and $A(\omega$.. The algebra $C^\infty(B(H_0))$, where solutions to the nonlinear equations are sought, is identified in the natural way with the algebra $\mathrm{Mat}_n(C^\infty(Z \times R^m))$ of matrix functions of the nth order $A\{k; x, y, \ldots\}$ that are infinitely differentiable with respect to $x, y, \ldots$. The extended space consists of vector functions

$$\vec{f}(z) = \sum_{k=1}^{n} f_k(z)\vec{e}_k$$

with the scalar product

$$\left(\vec{f}(z), \vec{g}(z)\right)_H = \mu^{-1} \sum_{k=1}^{n} \int f_k(z)\overline{g_k(z)}d\mu(z),$$

while the projection operators (3.4.1) are defined by the equality

$$\widehat{P}(\vec{f}(z)) = \sum_{i=1}^{n}(\sum_{k=1}^{n} \int p_{ik}(z) f_k(z) d\mu(z))\vec{e_i},$$

where the operator function matrix $\hat{p}(z)$ is denoted by $p(z) = (p_{ik}(z))$. Henceforth we shall assume that this matrix is a diagonal one:

$$p(z) = \mathrm{diag}(p_1(z), p_2,(z), \ldots, p_n(z)), \qquad \int p_k(z)d\mu(z) = 1,$$

whence

$$\widehat{P}(\vec{f}(z)) = \sum_{k=1}^{n} \left(\int p_k(z) f_k(z) d\mu \right) \vec{e_k}.$$

The operators $\hat{A} \in B(H)$ map the vectors

$$\vec{f} = \sum_{k=1}^{n} \nu_k \vec{e} \in H_0$$

into vector functions

$$\hat{A}(\vec{f}) = \sum_{i=1}^{n} \left(\sum_{k=1}^{n} a_{ik}(z) \nu_k \right) \vec{e_i} \in H,$$

($\sum_{i=1}^{n} a_{ik}(z)\vec{e_i}$ is the value of the operator $\hat{A}$ on the unit basis vector $\vec{e_k}$), whence it follows that on the subspace H_0 the equality $\hat{A}(\vec{f}) = \hat{a}(z)(\vec{f})$ is valid; here $\hat{a}(z) \in L^2_\mu(\Omega, B(H_0))$ stands for the decomposable operator with the matrix $a(z) = (a_{ik}(z))$. So the operators $\hat{A}\widehat{P}$ and $\widehat{P}\hat{A}\widehat{P}$ assume the form

$$\hat{A}\widehat{P} = \hat{a}(z)\widehat{P}, \qquad \widehat{P}\hat{A}\widehat{P} = \hat{a}\widehat{P}, \tag{4.0.1}$$

where $\hat{a}(z) \in L^2_\mu(\Omega, B(H_0))$ is a decomposable operator with the matrix $a(z) = (a_{ik}(z))$; $a \in B(H_0)$ and its matrix a equal $\int p(z)a(z)d\mu(z)$. If $\hat{A} \in C^\infty(B(H))$, then the matrices $a(z)$ and a depend, as can be easily seen, on the variables $x, y, \ldots$, too, and are infinitely differentiable with respect to these variables.

In what follows, matrices of decomposable operators will be denoted by the same letter without a cap.

§1 Examples of Solutions to Nonlinear Equations

In this section we shall consider particular cases of Theorems 3.6.1 and 3.6.2 and investigate the corresponding solutions of nonlinear equations.

1 *The KdV equation*

From formulas (4.0.1) that give the general form of the operators $\hat{A}\widehat{P}$, $\widehat{P}\hat{A}\widehat{P}$, it follows that $(\hat{\gamma} - iz)\widehat{P} = \hat{g}(z)\widehat{P}$, where the decomposable operator $\hat{g}(z)$ belongs to $L^2_\mu(\Omega, B(H_0))$. If the conditions of Theorem 3.6.1 are met, then the operator $\hat{g}(z)$ exists, and its matrix $g(z)$ is found from the equation

$$\begin{aligned} g(z) &+ \mathcal{E}(-2z)r_1(z)r_1^*(-\bar{z})\varphi(z)[\chi_2(z)m(z)g(-z) \\ &+ \int \frac{p^*(-\bar{z})p(z')}{i(z'+z)} g(z')d\mu(z')] = -\mathcal{E}(-2z)r_1(z)r_1^*(-\bar{z})p^*(-\bar{z})\varphi(z) \end{aligned} \tag{4.1.1}$$

that has a unique solution whose existence is ensured by Theorem 3.6.1.

In the capacity of $m(z)$, we take the matrix

$$m(z) = p^*(-\bar{z})\nu(z)p(-z), \qquad \nu^*(z) = \nu(\bar{z}), \qquad \chi_2(z)\nu(z) = \nu(z).$$

Introducing the notation

$$v(z) = p(z)g(z), \quad \rho(z) = \rho(z,x,t) = \mathcal{E}(-2z)p(z)r_1(z)r_1^*(-\bar{z})p^*(-\bar{z})\varphi(z),$$

we rewrite Eq. (4.1.1) in the form

$$v(z) + \rho(z)[\nu(z)v(-z) + \fint \frac{v(z')}{i(z'+z)}d\mu(z') + I] = 0. \tag{4.1.2}$$

According to Theorem 3.6.1, the matrices

$$u(x,t) = -2\partial_x \int v(z)d\mu(z) = -2\partial_x \int p(z)g(z)d\mu(z) \in C^\infty(x,t)$$

are self-adjoint solutions of the KdV equation

$$4\partial_t u + 4\lambda\partial_x u + \partial_x^3 u - 3\partial_x(u^2) = 0.$$

We introduce the following notation: $\Omega_R(\Omega_I)$ is the intersection of the support Ω of the measure $d\mu(z)$ with the real (imaginary) axis; $\chi_R(z), \chi_I(z)$ are characteristic functions of the sets Ω_R, Ω_I;

$$a_R(z) = \chi_R(z)a(z), \qquad a_I(z) = \chi_I(z)a(z).$$

Eq. (4.1.2) can be, obviously, decomposed into such a system of equations:

$$\begin{aligned}
&v_R(z) + \rho_R(z)[\nu_R(z)v_R(-z) + \fint \frac{v_R(z')}{i(z'+z)}d\mu(z') \\
&\quad + \int \frac{v_I(z')}{i(z'+z)}d\mu(z') + I] = 0 \qquad (z \in \Omega_R), \\
&v_I(z) + \rho_I(z)[\nu_I(z)v_I(-z) + \int v_R(z')i(z'+z)d\mu(z') \\
&\quad + \fint \frac{v_I(z')}{i(z'+z)}d\mu(z') + I] = 0 \qquad (z \in \Omega_I).
\end{aligned}$$

Under the conditions of Theorem 3.4.1, the measure $d\mu(z)$ on the set Ω_R coincides with the Lebesgue measure $(2\pi)^{-1}|dz|$, and, according to formulas of Plemelj–Sochocki,

$$\fint \frac{v_R(z')}{i(z'+z)}d\mu(z') = \int \frac{v_R(z')}{i(z'+z-i0)}d\mu(z') - \frac{1}{2}v_R(-z),$$

where

$$\int \frac{v_R(z')}{i(z'+z-i0)}d\mu(z') = \lim_{\varepsilon \to +0} \int \frac{v_R(z')}{i(z'+z-i\varepsilon)}d\mu(z').$$

For this reason, the obtained system of equations is equivalent to the following one:

$$v_R(z)+\rho_R(z)\Big\{(\nu_R(z)-\frac{1}{2})v_R(-z)+\int\frac{v_R(z')}{i(z'+z-i0)}d\mu(z')$$
$$+\int\frac{v_I(z')}{i(z'+z)}d\mu(z')+I\Big\}=0,\qquad (z\in\Omega_R),$$
$$v_I(z)+\rho_I(z)\{\nu_I(z)v_I(-z)+\int\frac{v_R(z')}{i(z'+z)}d\mu(z')$$
$$+\int\frac{v_I(z')}{i(z'+z)}d\mu(z')+I\}=0\qquad (z\in\Omega_I).$$

Let us consider two particular cases.

A). The set Ω_I consists of a finite number of points

$$i\kappa_1, i\kappa_2,\ldots,i\kappa_N\qquad (0<\kappa_1<\cdots<\kappa_N)$$

lying on the positive ray of the imaginary axis and $\nu_R(z)=\frac{1}{2}$. In this case, system (4.1.3) coincides, except for notational differences, with the equation of the inverse scattering problem, and the solutions of the KdV equation obtained from this system coincide with those found by the inverse problem method. These solutions decay rapidly when $x\to\pm\infty$. In the case $\Omega_R=\emptyset$, one obtains pure N-soliton solutions. For instance, one-soliton solutions have the form

$$u(x,t)=-4\kappa[\mathcal{E}(i\kappa)+\frac{m}{2\kappa}\mathcal{E}(-i\kappa)]^{-2}m,$$

where

$$m=p(i\kappa)r_1(i\kappa)r_1^*(i\kappa)p^*(i\kappa)$$

is an arbitrary constant nonnegative matrix,

$$\mathcal{E}(i\kappa)=\exp-\kappa\{x-(\kappa^2+\lambda)t\},$$

and κ is an arbitrary positive number. Therefore, in the appropriate basis they can be transformed into the diagonal form whose diagonal elements are scalar solitons differing from each other only by phases. In the general case, when $N\geqslant 2$, such decomposition into scalar solitons does not exist.

B). The set Ω_I consists of a finite number of intervals lying symmetrically on the imaginary axis (with respect to the origin), the set Ω_R is empty, and $\nu_I(z)=0$. In this case, system (4.1.3) is reduced to one equation,

$$v(z)+\rho(z)\{I+\int_{\Omega_I}\frac{v(z')}{i(z'+z)}d\mu(z')\}=0\qquad (z\in\Omega_I),\tag{4.1.4}$$

and all the conditions of Theorem 3.4.1 are met, if

$$d\mu(z)=\pi^{-1}|dz|=(i\pi)^{-1}dz\qquad (z\in\Omega_I),$$
$$p(z)=p_0(z)(w(z))^{-\frac{1}{2}},\qquad \sup_{z\in\Omega_I}\|p_0(z)\|_{H_0}<\infty,$$
$$\sup_{z\in\Omega_I}\|r(z)\|_{H_0}<\infty,\qquad \inf_{z\in\Omega_I}w(z)w(-z)>0,$$

where $w(z)$ is an arbitrary Muckenhoupt weight on the imaginary axis.

Thus, in our case,

$$\rho(z) = \varphi(z)\mathcal{E}(-z)^2 w(z)^{-1}\rho_1(z),$$

where

$$\rho_1(z) = p_0(z)r_1(z)r_1^*(-\bar{z})p_0^*(-\bar{z})$$

is an arbitrary bounded and positive (on the set Ω_I) matrix function, while Eq. (4.1.4) is evidently equivalent to the following equation,

$$\psi_1(z) - i\varphi(z)w_{(z)}^{-1}\rho_1(z)\mathcal{E}(-z)\{I + \int \frac{\chi_I(z')\mathcal{E}(-z')\psi_1(z')}{\pi i(z'+z)}dz\} = 0 \tag{4.1.5}$$

with respect to the matrix function

$$\psi_1(z) = \psi_1(z;x,t) = -i\mathcal{E}(z)v(z).$$

Let us set

$$\psi(z) = \psi(z;x,t) = \mathcal{E}(z)\{I + \int_{-i\infty}^{i\infty} \frac{\chi_I(z')\mathcal{E}(-z')\psi_1(z')}{\pi i(z'-z)}dz'\}$$

and clarify which properties this function possesses. It is clear that it is holomorphic in the entire complex plane with cuts along the intervals forming the set Ω_I and in the neigbourhood of the infinitely remote point:

$$\psi(z) = \exp iz(x + (z^2 - \lambda)t)\{I + \sum_{k=1}^{\infty} z^{-k}v_k\}, \tag{4.1.6}$$

where

$$\begin{aligned} v_k &= v_k(x,t) \\ &= \frac{i}{\pi}\int_{-i\infty}^{i\infty} \chi_I(z)z^{k-1}\mathcal{E}(-z)\psi_1(z)dz \\ &= \frac{1}{\pi}\int_{-i\infty}^{i\infty} \chi_I(z)z^{k-1}v(z)dz = i\int_{\Omega_I} z^{k-1}v(z)d\mu(z). \end{aligned}$$

From this equality it follows, in particular, that the desired solution to the KdV equation is given by the formula

$$u(x,t) = 2i\frac{d}{dx}v_1(x,t). \tag{4.1.7}$$

Denote by $\psi^+(z)$ (resp., $\psi^-(z)$) the left (resp., right) limit value of the function $\psi(z)$ on the intervals along which the cuts are made. According to the Plemelj–Sochocki formula,

$$\psi^{\pm}(z) = \mathcal{E}(z)\{I + \int_{-i\infty}^{i\infty} \frac{\chi_I(z')\mathcal{E}(-z')\psi_1(z')}{\pi i(z'-z)}dz' \pm \mathcal{E}(-z)\psi_1(z)\chi_I(z)\},$$

whence, employing Eq. (4.1.5), we find that

$$\psi_1(-z) - i\varphi(-z)w(-z)^{-1}\rho_1(-z)\{\psi^{\pm}(z) \mp \psi_1(z)\} = 0, \qquad z \in \Omega_I,$$

that is, on the cuts we have

$$\begin{gathered} i\varphi(-z)w(-z)^{-1}\rho_1(-z)\psi^{\pm}(z) = \psi_1(-z) \pm i\varphi(-z)w(-z)^{-1}\rho_1(-z)\psi_1(z), \\ i\varphi(z)w(z)^{-1}\rho_1(z)\psi^{\pm}(-z) = \psi_1(z) \pm i\varphi(z)w(z)^{-1}\rho_1(z)\psi_1(-z), \end{gathered}$$

Eliminating the matrix $\psi_1(z)$ from these equalities, we arrive at the following relations on the cuts:

$$\begin{aligned} & i\varphi(-z)w(-z)^{-1}\rho_1(-z)\psi^{\pm}(z) \\ &\quad = \psi_1(-z) \pm i\varphi(-z)w(-z)^{-1}\rho_1(-z)i\varphi(z)w(z)^{-1}\rho_1(z)\{\psi^{\pm}(-z) \mp \psi_1(-z)\} \\ &\quad = \mp\varphi(-z)\varphi(z)w(-z)^{-1}w(z)^{-1}\rho_1(-z)\rho_1(z)\psi^{\pm}(-z)\rho_1(z)\psi^{\pm}(-z) \\ &\qquad + (I + \varphi(-z)\varphi(z)w(-z)^{-1}w(z)^{-1}\rho_1(-z)\rho_1(z))\psi_1(-z) \\ &\qquad\quad = \pm w(-z)^{-1}w(z)^{-1}\rho_1(-z)\rho_1(z)\psi^{\pm}(-z) \\ &\qquad\qquad + (I - w(-z)^{-1}w(z)^{-1}\rho_1(-z)\rho_1(z))\psi^{(-z)} \end{aligned}$$

since for the function $\varphi(z) = \chi_0(z) - \chi_+(z) + \chi_-(z)$, the identity $\varphi(z)\varphi(-z) \equiv -1$ holds on the set Ω_I.

In particular, if

$$w(-z)^{-1}w(z)^{-1}\rho_1(-z)\rho_1(z) \equiv I \qquad (z \in \Omega_I), \tag{4.1.8}$$

then the limit values of the function $\psi(z)$ on the cuts are related by the following expression,

$$i\varphi(-z)w(-z)^{-1}\rho_1(-z)\psi^{\pm}(z) = \pm\psi^{\pm}(-z). \tag{4.1.9}$$

Thus, if relation (4.1.8) holds, then in order to find the solution of the KdV equation it suffices to construct the matrix function $\psi(z)$ which

1) is holomorphic throughout the plane cut along the intervals of the set Ω_I;

2) is representable in the form (4.1.6) in the neighbourhood of the infinitely remote point;

3) satisfies boundary conditions (4.1.9) on the cuts. Existence and uniqueness[1] of the matrix function is ensured by Theorem 3.6.1.

Denote now the intervals on the imaginary axis that make the set Ω_I by

$$(ia_k, ib_k) = \Delta_k, \qquad (-ib_k, -ia_k) = \bar{\Delta}_k$$
$$(0 < a_1 < b_1 < a_2 < b_2 < \cdots < a_N < b_N).$$

[1]To be exact, uniqueness of the decomposable operator $\hat{\psi}(z)$, since its matrix depends on the choice of a basis in the space H_0.

The polynomial

$$p(z) = (-1)^N \prod_{k=1}^{N} (z - ia_k)(z - ib_k)$$

assumes real values on the imaginary axis with $P(iy) > 0$ if $iy \notin \cup_{k=1}^{N} \Delta_k$, and with $P(iy) < 0$ if $iy \in \cup_{k=1}^{N} \Delta_k$. Therefore, in the plane cut along the intervals $\Delta_k (k = 1, 2, \dots, N)$, there exists a holomorphic branch of the radical $R(z) = \sqrt{P(z)}$ which is positive on the semiaxis $(ib_N, +i\infty)$. In the neighbourhood of the infinitely remote point, this branch may be expanded into a series

$$R(z) = (-i)^N z^N (1 + \frac{\alpha}{z} + \dots),$$

with the branch limit values on the cuts satisfying the relations

$$R(z)^+ = -R(z)^-, \qquad iR(z)^+(-1)^{N-k} > 0, \qquad z \in \Delta_k,$$

whence it follows that the limit value of the ratio $R(z)/R(-z)$ possesses the properties

$$\left(\frac{R(z)}{R(-z)}\right)^+ = -\left(\frac{R(z)}{R(-z)}\right)^-, \quad i\varphi(z)\left(\frac{R(z)}{R(-z)}\right)^+ (-1)^k > 0, \quad z \in \Delta_k \cup \bar{\Delta}_k,$$

where $\varphi(z) = -\chi_+(z) + \chi_-(z)$. Further, it is easy to check that if the real values $c_k (1 \leqslant k \leqslant N)$ satisfy the inequalities

$$0 < |c_1| < a_1 < b_1 < |c_2| < \dots < b_{N-1} < |c_N| < a_N,$$

then the rational fraction

$$\frac{P_1(z)}{P_1(-z)}, \qquad P_1(z) = \prod_{k=1}^{N} (z - ic_k)$$

assumes on the sets $\Delta_k \cup \bar{\Delta}_k$ real values of different signs and

$$\frac{P_1(-z)}{P_1(z)}(-1)^k > 0, \qquad z \in \Delta_k \cup \bar{\Delta}_k.$$

Comparing the relations obtained, we see that the inequalities

$$\pm i\varphi(z)\frac{P_1(-z)}{P_1(z)}\left(\frac{R(z)}{R(-z)}\right)^\pm > 0$$

are satisfied on the entire set Ω_I, that is

$$\pm i\varphi(z)\frac{P_1(-z)}{P_1(z)}\left(\frac{R(z)}{R(-z)}\right)^\pm = \left|\frac{P_1(-z)}{P_1(z)}\right|\left|\frac{R(z)}{R(-z)}\right|. \tag{4.1.10}$$

Consider now a matrix function

$$\tilde{\psi}(z) = (-i)^{-N} \frac{R(z)\psi(z)}{P_1(z)}.$$

This function is meromorphic in the plane cut along the set Ω_I and has only simple poles in the points ic_k $(1 \leqslant k \leqslant N)$; when $|z| \to \infty$,

$$\tilde{\psi}(z) = \exp iz(x + (z^2 - \lambda)t)(I + \sum_{k=1}^{\infty} z^{-k}\tilde{v}_k),$$

where the coefficient $\tilde{v}_1$ differs from v_1 by a constant term. Therefore formula (4.1.7) is still valid:

$$u(x,t) = 2i\frac{d}{dx}\tilde{v}_1(x,t).$$

According to Eq. (4.1.9), limit values of the function $\tilde{\psi}(z)$ on the set Ω_I are related in the following way:

$$i\varphi(-z)w(-z)^{-1}\rho_1(-z)\frac{P_1(z)}{R(z)^{\pm}}\tilde{\psi}^{\pm}(z) = \pm\frac{P_1(-z)}{R(-z)^{\pm}}\tilde{\psi}^{\pm}(-z).$$

Multiplying both parts of these equalities by $i\varphi(z)R(z)^{\pm}P_1(z)^{-1}$, we find that

$$w(-z)^{-1}\rho_1(-z)\tilde{\psi}^{\pm}(z) = \pm i\varphi(z)\left(\frac{R(z)}{R(-z)}\right)^{\pm}\frac{P_1(-z)}{P_1(z)}\tilde{\psi}^{\pm}(-z)$$

and, from Eq. (4.1.10),

$$w(-z)^{-1}\rho_1(-z) = \tilde{\psi}(z)\left|\frac{R(z)}{R(-z)}\right|\left|\frac{P_1(-z)}{P_1(z)}\right|\tilde{\psi}^{\pm}(-z). \tag{4.1.11}$$

Since the function

$$\frac{|R(z)|}{|R(-z)|}$$

is a Muckenhoupt weight, the function

$$\frac{|P_1(z)|}{|P_1(-z)|}$$

is bounded on the set Ω_I, and without violating the conditions of Theorem 3.6.1, we may set

$$w(z) = \left|\frac{R(z)}{R(-z)}\right|, \rho_1(z) = \left|\frac{P_1(z)}{P_1(-z)}\right|\rho_0(z),$$

where $\rho_0(z)$ is an arbitrary positive matrix function satisfying the condition $\rho_0(z)\rho_0(-z) = I$. Then equalities (4.1.11) assume the form

$$\rho_0(-z)\tilde{\psi}^{\pm}(z) = \tilde{\psi}^{\pm}(-z).$$

In the scalar case ($n = 1$), by setting $\rho_0(-z) \equiv 1$, we find that

$$\tilde{\psi}^{\pm}(z) = \tilde{\psi}^{\pm}(-z). \tag{4.1.12}$$

The function z^2 conformly maps the complex plane with cuts along the set Ω_I onto a two-sheeted Riemann surface with a branching point at zero and cuts $\tilde{\Delta}_k$ ($\tilde{\tilde{\Delta}}_k$) along segments $(-b_k^2, -a_k^2)$ of the upper (lower) sheet; points ic_k are mapped into points $\tilde{c}_k = -c_k^2$ of the upper sheet if $c_k > 0$, and into those of the lower sheet if $c_k < 0$. On the Riemann surface with the above-mentioned cuts, the function $\tilde{\psi}(\sqrt{z})$ is meromorphic in any finite domain, it has simple poles at points $\tilde{c}_k$, in the neighbourhood of the infinitely remote point

$$\tilde{\psi}(\sqrt{z}) = \exp i\sqrt{z}(x + (z - \lambda)t)(1 + \sum_{k=1}^{\infty}(\sqrt{z})^{-k}\tilde{v}_k, \tag{4.1.13}$$

and, according to Eq. (4.1.12), along the cuts, it satisfies the boundary conditions

$$\tilde{\psi}^{\pm}(\sqrt{z}, +) = \tilde{\psi}^{\mp}(\sqrt{z}, -). \tag{4.1.12'}$$

Here $\tilde{\psi}^{+}$ ($\tilde{\psi}^{-}$) stands for the limit value of the function $\tilde{\psi}(\sqrt{z})$ on upper (lower) edges of the cuts, and $(\sqrt{z}, +), (\sqrt{z}, -)$ denote points of the upper and lower sheets, respectively. Connecting the upper (lower) edge of the cut $\tilde{\Delta}_k$ with the lower (upper) edge of the cut $\tilde{\tilde{\Delta}}_k$, we obtain a new Riemann surface, namely that of the function

$$\sqrt{z(z + a_1^2)(z + b_1^2)\ldots(z + a_N^2)(z + b_N^2)}.$$

It follows from equalities (4.1.12′) that the function $\tilde{\psi}(\sqrt{z})$ is meromorphic everywhere on this surface, except in the neighbourhood of the infinitely remote point, where it expands into series (4.1.13), and has but simple poles at the points $\tilde{c}_k$ ($1 \leqslant k \leqslant N$). A function possessing such properties is said to be a Baker–Akhiezer one.

Askiezer [1] showed that finding such a function is reduced to solving a certain Jacobi inversion problem. It enabled Its and Matveew [12] to find an explicit (via the Riemann θ-function) expression for it, and hence, for the function

$$u(x, t) = 2_i\partial_x^2\tilde{v}_1(x, t).$$

In the fundamental work by Novikov [21] (see also Lax [15], [31]) it was shown that the obtained solution $u(x, t)$ is a quasi-periodic function, while the spectrum of the operator $-d^2/dx^2 + u(x, t)$ ($-\infty < x < \infty$) consists of the finite number of zones $(-b_N^2, -a_N^2), \ldots, (-b_1^2, -a_1^2), (0, \infty)$. For a detailed description of these results, see monograph [4].

Thus, the class of finite-zone solutions of the scalar KdV equation is obtained with the following choice of parameters in Theorem 3.6.1:

1) $\Omega = \Omega_I = \cup_{k=1}^{N}(\Delta_k \cup \tilde{\Delta}_k)$, $\quad d\mu(z) = (i\pi)^{-1}dz$,

2) $m(z) = 0$,

3) $w(z) = |R(z)||R(-z)|^{-1}$,
4) $p_0(z) - (z) = |P_1(z)||P_1(-z)|^{-1}$.

Note that beside the above properties the essential one is the requirement (4.1.8), for it is this requirement that enables reduction of the integral equation (4.1.5) in solving a corresponding Hilbert–Riemann probalem on the Riemann surface of the function

$$\sqrt{z(z+a_1^2)(z+b_1^2)\dots(z+b_N^2)}.$$

2 *MKdV, NLSH, SG Equations*

Assume that conditions of Theorem 3.6.2 are met. Then there exists the logarithmic derivative $\hat{\gamma} = \widehat{\Gamma}^{-1}\partial_x\widehat{\Gamma}$,

$$\hat{\gamma} - \widehat{N}(z) = \widehat{T}^{-1}\widehat{R}(z)\hat{p}^*(z)\widehat{B}\widehat{P} = (\hat{\gamma}(z) - \widehat{N}(z))\widehat{P} = (\hat{\gamma}(z) - \widehat{N}(z))\widehat{P},$$

and the matrix $\gamma(z) - N(z)$ is found from the equation

$$(\gamma(z) - N(z)) + R(z)p^*(z)\{\widehat{L}_0(\gamma(z) - N(z)) + B\} = 0,$$

which is equivalent to

$$g(z) + \mathcal{E}^{-1}(z)\rho(z)\mathcal{E}^{-1}(z)^*\{\widehat{L}_0(g(z)) + B\} = 0 \tag{4.1.14}$$

in the matrices

$$g(z) = p(z)(\gamma(z) - N(z)),$$

where

$$B = Q_1 - Q_2, \qquad r(z) = r^*(z) = Q_1 r(z) Q_2 + Q_2 r(z) Q_1,$$
$$\rho(z) = p(z)r(z)p^*(z), \qquad Q_i p(z) = p(z)Q_i.$$

We consider in detail the most interesting case when the space H_0 is two-dimensional. In this case,

$$B = \begin{pmatrix} 1 & 0 \\ 0 & -1 \end{pmatrix}, \qquad \nu(z) = \begin{pmatrix} \nu_1(z) & 0 \\ 0 & \nu_2(z) \end{pmatrix}, \qquad \nu_1(z) = \overline{\nu_1(\bar{z})},$$
$$\mathcal{E}(z)^{-1}\rho(z)\mathcal{E}^*(z)^{-1} = \begin{pmatrix} 0 & \rho(z)e^{-2\theta(z)} \\ \overline{\rho(z)}e^{-2\overline{\theta(z)}} & 0 \end{pmatrix}$$

and Eq. (4.1.14) decomposes into such a system of equations in elements $g_{ik}(z)$ of the matrix $g(z)$:

$$\begin{aligned} &g_{11}(z) + e^{-2\theta(z)}\rho(z)(\widehat{\Lambda}_2 + \nu_2(z)\hat{\pi}_2)g_{21}(z) = 0, \\ &g_{21}(z) + e^{-2\overline{\theta(z)}}\overline{\rho(z)}\{(\widehat{\Lambda}_1 + \nu_1(z)\hat{\pi}_2)g_{11}(z) + 1\} = 0, \\ &g_{12}(z) + e^{-2\theta(z)}\rho(z)\{(\widehat{\Lambda}_2 + \nu_2(z)\hat{\pi}_2)g_{22}(z) - 1\} = 0, \\ &g_{22}(z) + e^{-2\overline{\theta(z)}}\overline{\rho(z)}(\widehat{\Lambda}_1 + \nu_1(z)\hat{\pi}_2)g_{12} = 0. \end{aligned} \tag{4.1.14'}$$

(We recall that the function $\theta(z)$ is defined by Eq. (3.4.11).) The constraints imposed on the matrix $\nu(z)$ by the conditions of Theorem 3.6.2 are met if

$$\nu_1(z) = -\overline{\nu_2(z)} = \nu(z) = \overline{\nu(\bar{z})}, \qquad \inf_{-\infty<z<\infty} \nu(z) \geqslant \frac{1}{2}.$$

Assuming that these conditions are satisfied, we carry out complex conjugation in the second and fourth equation of system (4.1.14′). Since $\bar{\Lambda}_1 = -\Lambda_2$, it will result in the same equation for the vector functions

$$\begin{pmatrix} g_{21}(z) \\ -g_{11}(z) \end{pmatrix}, \qquad \begin{pmatrix} -\overline{g_{12}(z)} \\ -\overline{g_{22}(z)} \end{pmatrix},$$

namely,

$$C\hat{I}(\vec{g}(z)) + e^{-2\theta(z)}\rho(z)\left\{(\hat{\Lambda}_2 - \overline{\nu(z)}\hat{\pi}_2)\vec{g}(z) + \begin{pmatrix} 0 \\ 1 \end{pmatrix}\right\} = 0, \tag{4.1.15}$$

where $\hat{I}$ is a conjugation operator ($\hat{I}(\varphi(z)) = \overline{\varphi(z)}$, while

$$C = \begin{pmatrix} 0 & -1 \\ 1 & 0 \end{pmatrix}.$$

That means that the vectors coincide when the desired matrix has the form

$$g(z) = p(z)(\gamma(z) - N(z)) = \begin{pmatrix} -\overline{g_2(z)} & -\overline{g_1(z)} \\ g_1(z) & -g_2(z) \end{pmatrix}, \tag{4.1.16}$$

and the functions $g_1(z)$, $g_2(z)$ are found from Eq. (4.1.15), which is satisfied by the vector

$$g(z) = \begin{pmatrix} g_1(z) \\ g_2(z) \end{pmatrix} = \begin{pmatrix} g_{21}(z) \\ -g_{11}(z) \end{pmatrix} = \begin{pmatrix} -\overline{g_{12}(z)} \\ -\overline{g_{22}(z)} \end{pmatrix}.$$

Combining the results of §III.5 and Eq. (4.1.16), we conclude that

1) the function

$$u_1 = u_1(x,t) = \int g_1(z)d\mu(z)$$

solves the equation

$$-2i\partial_t u_1 + \partial_x^2 u_1 + 8|u_1|^2 u_1 = 0$$

if $\theta(z) = iz(x + zt)$, or the equation

$$4\partial_t u_1 + \partial_x^3 u_1 + 16|u_1|^2 \partial_x u_1 = 0$$

if $\theta(z) = iz(x + z^2 t)$;

2) the matrix

$$v = \begin{pmatrix} v_2 & v_1 \\ -\overline{v_1} & \overline{v_2} \end{pmatrix},$$

$$v_1 = i\int z^{-1}\overline{g_1(z)}d\mu(z),$$

$$v_2 = 1 + i\int z^{-1}\overline{g_2(z)}d\mu(z)$$

is unitary and satisfies the equation

$$\partial_y(v^*\partial_x v) = v^*BvB - Bv^*Bv \tag{4.1.17}$$

if $\theta(z) = i(zx - z^{-1}y)$, and the matrix $S = vBv^* = S^{-1}$ is a self-adjoint and unitary solution of the equation

$$-4i\partial_t S = [S, \partial_x^2 S]$$

if $\theta(z) = iz(x + zt)$.

If, in addition,

$$\overline{\nu(z)} = \nu(-\bar{z}), \qquad \overline{r(z)} = -r(-\bar{z}), \qquad d\mu(z) = d\mu(-\bar{z}),$$

then $v_2 = \bar{v}_2$, $v_1 = -\bar{v}_1$, the matrix v has the form

$$v = \begin{pmatrix} \cos u & i\sin u \\ i\sin u & \cos u \end{pmatrix}, \qquad u = \arg(v_2 + v_1),$$

and Eq. (4.1.17) is reduced to an SG equation

$$\partial_y\partial_x u = 2\sin 2u.$$

Using the above notation, one can split Eq. (4.1.15) into the following system:

$$\begin{aligned}
C\hat{I}(\vec{g}_R(z)) + e^{-2\theta(z)}\rho(z)\Big\{&(\overline{-\nu_R(z)} + \frac{1}{2})\vec{g}_R(z) \\
&+ \int \frac{\vec{g}_R(\bar{z}')}{i(z' - z + io)}\, d\mu(z') \\
&+ \int \frac{\vec{g}_I(\bar{z}')}{i(\bar{z}' - z)}\, d\mu(z') + \begin{pmatrix} 0 \\ 1 \end{pmatrix}\Big\} = 0, \qquad z \in \Omega_R \\
C\hat{I}(\vec{g}_I(z)) + e^{-2\theta(z)}\rho(z)\Big\{&\overline{-\nu_I(z)}g_I(z) \\
&+ \int \frac{g_R(z')}{i(z' - z)}\, d\mu(z') \\
&+ \int \frac{g_I(z')}{i(\bar{z}' - z)}\, d\mu(z') + \begin{pmatrix} 0 \\ 1 \end{pmatrix}\Big\} = 0
\end{aligned} \tag{4.1.18}$$

in the vector functions

$$\vec{g}_R(z) = \chi_0(z)\vec{g}(z), \qquad \vec{g}_I(z) = (\chi_+(z) + \chi_-(z))\vec{g}(z).$$

Let us consider two particular cases.

A) The measure $d\mu(z)$ is concentrated in the bounded upper half-plane $w(z) = 1$ and $\nu_R(z) = \frac{1}{2}$. Since $\nu_I(z) = 0$ in this case, all conditions of Theorem 3.6.2. are met if $\chi_0(z)d\mu(z) = (2\pi)^{-1}|dz|$ and the measure $\chi_+(z)d\mu(z)$ satisfy the standard Carleson condition. For instance, if the measure $\chi_+(z)d\mu(z)$ is concentrated in a finite number of points, then system (4.1.18) coincides up to notational differences with the system of equations of the inverse problem of scattering theory. In this case, we arrive at the same class of solutions that the inverse problem method gives as well.

B) The measure is concentrated on the finite number of segments

$$\Delta_k = (a_k, b_k), \quad \bar{\Delta}_k = (\bar{b}_k, \bar{a}_k)(\operatorname{Re} a_k = \operatorname{Re} b_k, \operatorname{Im} b_k > \operatorname{Im} a_k \geqslant 0, \quad 1 \leqslant k \leqslant N)$$

of vertical straights. The conditions of Theorem 3.6.2 are met if

$$d\mu(z) = \pi^{-1}|dz| = (i\pi)^{-1}dz, \qquad \rho(z) = \rho_0(z)w(z)^{-1}, \qquad \nu(z) = 0,$$
$$\inf_z w(z)w(\bar{z}) > 0, \qquad \rho_0(z)\overline{\rho_0(\bar{z})} > 0,$$

where $\rho_0(z)$ is a smooth function on the set $\Omega = \cup_{k=1}^N(\Delta_k \cup \bar{\Delta}_k)$, and $w(z)$ is the Muckenhoupt weight on the vertical straights. In this case, system (4.1.8) is reduced to a single vector equation

$$\begin{pmatrix} \overline{-g_2(\bar{z})} \\ g_1(\bar{z}) \end{pmatrix} + e^{-2\theta(z)}\rho(z)\left\{\begin{pmatrix} 0 \\ 1 \end{pmatrix} + \int\limits_\Omega \frac{1}{i(\bar{z}' - z)}\begin{pmatrix} g_1(\bar{z}') \\ g_2(\bar{z}') \end{pmatrix} d\mu(z')\right\} = 0,$$

which can be conveniently written as a system of two scalar equations

$$(4.1.19) \qquad \begin{aligned} &\overline{-g_2(\bar{z})} + e^{-2\theta(z)}\rho(z)\int\limits_\Omega \frac{g_1(\bar{z}')}{i(z' - z)}d\mu(z') = 0, \\ &g_1(\bar{z}) + e^{2\theta(z)}\overline{\rho(\bar{z})}\{1 + \int\limits_\Omega \frac{\overline{-g_2(z')}}{i(z' - z)}d\mu(z')\} = 0. \end{aligned}$$

(We have used equalities $d\mu(z) = d\mu(\bar{z})$, $\overline{\theta(\bar{z})} = -\theta(z)$).

Let us consider functions

$$\varphi_1(z) = \int\limits_\Omega \frac{g_1(\bar{z}')}{i(z' - z)}d\mu(z'), \qquad \varphi_2(z) = 1 + \int\limits_\Omega \frac{\overline{-g_2(z')}}{i(z' - z)}d\mu(z'),$$

which are holomorphic in the entire plane with cuts along segments $\Delta_k, \bar{\Delta}_k (1 \leqslant k \leqslant N)$.

In the neighbourhood of the infinitely remote point

$$\varphi_1(z) = i\sum_{k=1}^{\infty} z^{-k}u_k(x,t),$$

where

$$u_1(x,t) = \int\limits_\Omega g_1(\bar{z})d\mu(z) = \int\limits_\Omega g_1(z)d\mu(z).$$

Besides, if the measure support does not contain zero, then these functions are holomorphic in the neighbourhood of zero and

$$\varphi_1(0) = -i \int_\Omega z^{-1} g_1(\bar{z}) d\mu(z) = \overline{(i \int_\Omega z^{-1} \overline{g_1(z)} d\mu(z))},$$

$$\varphi_2(0) = 1 + i \int_\Omega z^{-1} \overline{g_2(z)} d\mu(z).$$

So, in order to obtain solutions of the considered nonlinear equations, one needs to find the functions $\varphi_1(z)$, $\varphi_2(z)$ and calculate $\lim_{z\to\infty} z\varphi_1(z)$, $\varphi_1(0)$, $\varphi_2(0)$.

From the equality $d\mu(z) = (i\pi)^{-1}dz$ and Plemelj–Sochocki formulas, it follows that the functions $\varphi_1(z)$, $\varphi_2(z)$ have such limit values on the cuts:

$$\varphi_1^{\pm}(z) = \fint_\Omega \frac{g_1(\bar{z}')}{i(z'-z)} d\mu(z') \mp i g_1(\bar{z}),$$

$$\varphi_2^{\pm}(z) = 1 + \fint_\Omega \frac{-\overline{g_2(\bar{z}')}}{i(z'-z)} d\mu(z') \pm i\overline{g_2(z)}.$$

Employing these equalities and Eq. (4.1.19), we obtain the relations

$$\overline{-g_2(z)} + e^{-2\theta(z)} \rho(z) \{\varphi_1^{\pm}(z) \pm i g_1(\bar{z})\} = 0,$$
$$g_1(\bar{z}) + e^{2\theta(z)} \overline{\rho(\bar{z})} \{\varphi_2^{\mp}(z) \pm i\overline{g_2(z)}\} = 0.$$

We eliminate from them the function $\overline{g_2(z)}$ and arrive at the following boundary condition on the cuts:

$$g_1(\bar{z})\{1 - \rho(z)\overline{\rho(\bar{z})}\} + e^{2\theta(z)} \overline{\rho(\bar{z})} \varphi_2^{\mp}(z) \pm i\rho(z)\overline{\rho(\bar{z})} \varphi_1^{\pm}(z) = 0.$$

In particular, if

$$\rho(z)\overline{\rho(\bar{z})} \equiv 1, \tag{4.1.20}$$

then

$$\varphi_1^{\pm}(z) = \pm i e^{2\theta(z)} \overline{\rho(\bar{z})} \varphi_2^{\mp}(z), \tag{4.1.20'}$$

which enables us to reduce determination of the functions to solving the Riemann boundary problem on the Riemann surface of F of the function

$$\sqrt{\prod_{k=1}^{N} (z - a_k)(z - \bar{a}_k)(z - b_k)(z - \bar{b}_k)}.$$

To this end, we shall construct the surface F as a two-sheeted covering surface of the z-plane: namely, we shall take two samples F^+, F^- of the z-plane, cut along the segments Δ_k and $\bar{\Delta}_k (1 \leqslant k \leqslant N)$, place F^+ over F^- and connect the cut edges in criss-cross fashion. We denote $\mathcal{L}$ the loop formed by the connected cut edges and oriented as follows: the left cut edges of the F^+ plane must be passed upwards and the right ones downwards. On the surface F, this loop is a closed analytic curve consisting of $2N$ connected components. In the plane cut along the segments Δ_k $(1 \leqslant k \leqslant N)$, we shall select a holomorphic branch of the radical

$$R(z) = \sqrt{\prod_{k=1}^{N}(z-a_k)(z-b_k)}$$

which in the neighbourhood of the infinitely remote point expands into a series

$$R(z) = z^N(1 + z^{-1}\alpha_1 + z^{-2}\alpha_2 + \ldots).$$

Obviously the function $\overline{R(\bar{z})}$ is holomorphic in the plane cut along the segments $\bar{\Delta}_k$ $(1 \leqslant k \leqslant N)$,

$$\overline{R(\bar{z})} = z^N(1 + z^{-1}\bar{\alpha}_1 + z^{-2}\bar{\alpha}_2 + \ldots) \qquad (z \to \infty),$$

and the limit values of the functions $R(z)$, $\overline{R(\bar{z})}$ on all segments Δ_k, $\bar{\Delta}_k$ of the measure support are related in the following manner

$$\left(\frac{R(z)}{\overline{R(\bar{z})}}\right)^{\pm} = \pm\left(\frac{R(z)}{\overline{R(\bar{z})}}\right)^{+}, \qquad \frac{\overline{R(\bar{z})}^{\pm}}{\overline{R(\bar{z})}^{\mp}} = \delta(z), \tag{4.1.21}$$

where

$$\delta(z) = \chi_+(z) - \chi_-(z) = \begin{cases} 1 & z \in \Delta_k \\ -1 & z \in \bar{\Delta}_k \end{cases}.$$

Let us remove the loop $\mathcal{L}$ from the surface F, and on the remaining open set $F \backslash \mathcal{L}$ we consider a function

$$\Phi(z) = \begin{cases} \overline{R(\bar{z})}\varphi_2(z) & z \in F^+ \\ R(z)\varphi_1(z) & z \in F^- \end{cases}$$

The function is meromporphic on the set $F\backslash\mathcal{L}$ and has only two poles: one of multiplicity N at the point ∞^+ of the upper sheet F^+, another one of multiplicity $N-1$ at the point ∞^- of the lower sheet F^-. From Eqs. (4.1.20s′) and (4.1.21) it follows that on the loop $\mathcal{L}$ it satsifies the boundary condition

$$\Phi^+(z) = \delta(z)e^{2\theta(z)}\left\{i\overline{\rho(\bar{z})}\left(\frac{R(z)}{\overline{R(\bar{z})}}\right)^{+}\right\}\Phi^-(z) \qquad (z \in \mathcal{L}).$$

Now, we choose the parameters $w(z)$ and $\rho_0(z)$:

$$w(z) = \left|\frac{\overline{R(\bar{z})}}{R(z)}\right|, \qquad \rho_0(z) = i\left(\frac{R(z)}{\overline{R(\bar{z})}}\right)^{+}\left|\frac{R(z)}{\overline{R(\bar{z})}}\right|^{-1} q_0(z), \qquad q_0(z)\overline{q_0(\bar{z})} = 1,$$

where $q_0(z)$ is an arbitrary smooth function. Such a choice of the parameters ensures that all conditions of Theorem 3.6.2 and Eq. (4.1.20) are satisfied, while the boundary condition for the function $\Phi(z)$ assumes the form

$$\Phi^+(z) = \delta(z)q_0(z)e^{2\theta(z)}\Phi^-(z) \qquad (z \in \mathcal{L}). \tag{4.1.22}$$

Therefore, the function $\Phi(z)$ is a solution of the Riemann boundary problem (4.1.22) with a divisor of poles $(\infty^+)^{-N}$, $(\infty^-)^{-N}$ normalized by the condition

$$\lim_{z \longrightarrow \infty^+} z^N \Phi(z) = 1.$$

Existence and uniqueness of the solution of this boundary problem are guaranteed by Theorem 3.6.2. It is known [9] that its solution is expressed via the Riemann θ-function. Explicity formulas for obtaining the corresponding solutions to nonlinear equations were first obtained by Its, Kotlyarov and Kozel ([10], [11], [13]).

Thus, we see that in the case considered one gets finite-zone solutions of nonlinear equations. This solution class is determined by the parameters that possess the following properties:

$$\nu(z) = 0, \qquad \rho(z)\overline{\rho(\bar{z})} = 1.$$

We conclude by showing how the boundary problem (4.1.22) is connected with Baker–Akhiezer functions. To this end, we consider a function

$$\psi(z) = \begin{cases} \Phi(z)\Phi_0(z)^{-1}e^{\theta(z)}, & z \in F^+ \\ \Phi(z)\Phi_0(z)^{-1}e^{-\theta(z)}, & z \in F^- \end{cases}$$

where $\Phi_0(z)$ is a solution of the boundary problem (4.1.22) at $x = t = 0$. It follows from definitions of the functions $\Phi(z)$, $\Phi_0(z)$, that the function $\Phi(z)\Phi_0(z)^{-1}$ is meromorphic in the vicinity of infinitely remote points ∞^-, ∞^+ and expands there into the following power series

$$\Phi(z)\Phi_0(z)^{-1} = \begin{cases} u(x,t)u(0,0)^{-1} + \sum_{k=1}^{\infty} z^{-k} d_k^-(x,t) \\ 1 + \sum_{k=1}^{\infty} z^{-k} d_k^+(x,t) \end{cases}$$

while Eqs. (4.1.20) and (4.1.22) imply that on the loop $\mathcal{L}$, $\psi^+(z) = \psi^-(z)$. Thus, the function $\psi(z)$ is meromorphic on the entire surface F, except at the points ∞^+, ∞^- where it has essential singularities, the function poles lie in zeros of the function $\Phi_0(z)$ and do not depend on x, t. Note, finally, that Eq. (4.1.22) implies that the function

$$\mathcal{G}(z) = \begin{cases} 2\Phi_0(z)\overline{\Phi_0(\bar{z})} & z \in F^+ \\ -2\Phi_0(z)\overline{\Phi_0(\bar{z})} & z \in F^- \end{cases}$$

is meromorphic on the entire surface F, and the divisor of its poles is equal to $(\infty^+)^{2N}$, $(\infty^-)^{2N-2}$. Hence, $\mathcal{G}(z) = R(z)\overline{R(\bar{z})} + f(z)$, where $f(z)$ is a real polynomial $\sum_{k=1}^{2N} z^k f_k$ and $f_{2N} = 1$, $f_{2N-1} = (\alpha_1 + \bar{\alpha}_1)$. Since

$$(R(z)\overline{R(\bar{z})} + f(z))(R(z)\overline{R(\bar{z})} - f(z))$$

is a real polynomial of power $4N-2$, zeroes of the function $\mathcal{G}(z)$ coincide with those of the polynomial. Thus, the function $\Phi_0(z)$ has $2N-1$ zeroes $\mu_1, \mu_2, \ldots, \mu_{2N-1}$, their number being equal to the kind of surface F, while the function $\psi(z)$ is a Baker–Akhiezer one.

§2 Connection with Inverse Problems of Spectral Analysis

The solution method for nonlinear equations described in previous chapters used no variants of inverse problems of spectral analysis. It is clear, however, that such a connection with the inverse problems must exist. The present section is devoted to elucidation of this connection.

LEMMA 4.2.1. *Let the invertible operator $\widehat{\Gamma} \in C^\infty(B(H))$ satisfy the following equations*

$$\partial_x^2 \widehat{\Gamma} = \hat{A}\widehat{\Gamma}, \qquad \partial_x \widehat{\Gamma}(I-\widehat{P}) = \widehat{\Gamma}\widehat{N}(I-\widehat{P}), \tag{4.2.1}$$

where $\hat{A}, \widehat{P} \in B(H)$ and the subset $H_0 \subset H$ is the domain of values of the projection operator $\widehat{P} = \widehat{P}^2$. Then the operator

$$\widehat{Y}(N,x) = \widehat{Y} = e^{\widehat{N}x}(\hat{\gamma} - \widehat{N})\widehat{P} \qquad (\hat{\gamma} = \widehat{\Gamma}^{-1}\partial_x\widehat{\Gamma})$$

satisfies the Sturm–Liouville equation

$$\partial_x^2 \widehat{Y} - \widehat{Y}\hat{u}(x) - \widehat{N}^2\widehat{Y} = 0,$$

with the potential

$$\hat{u}(x) = -2\partial_x \widehat{P}\hat{\gamma}\widehat{P} \in C^\infty(B(H_0)),$$

while the operator $(\hat{\gamma} - \widehat{N})\widehat{P} = e^{-\widehat{N}x}\widehat{Y}$ satisfies the equation

$$\partial_x^2\{e^{-\widehat{N}x}\widehat{Y}\} - \{e^{-\widehat{N}x}\widehat{Y}\}\hat{u}(x) + 2\widehat{N}\partial_x\{e^{-\widehat{N}x}\widehat{Y}\} = 0.$$

PROOF: It was already shown that Eq. (4.2.1) implies such equations for the logarithmic derivative

$$\partial_x^2\hat{\gamma} + 2\hat{\gamma}\partial_x\hat{\gamma} = 0, \qquad \hat{\gamma}(I-\widehat{P}) = \widehat{N}(I-\widehat{P}). \tag{4.2.2}$$

Hence,

$$\hat{\gamma} = \widehat{N} + (\hat{\gamma} - \widehat{N})\widehat{P} = \widehat{N} + e^{-\widehat{N}x}\widehat{Y},$$

and since

$$\partial_x\hat{\gamma} = \partial_x\hat{\gamma}\widehat{P}, \qquad \widehat{Y} = \widehat{Y}\widehat{P},$$

then

$$\begin{aligned}
\partial_x^2\hat{\gamma} &= \partial_x^2(e^{-\widehat{N}x}\widehat{Y}) = e^{-\widehat{N}x}\{\widehat{N}^2\widehat{Y} - 2\widehat{N}\partial_x\widehat{Y} + \partial_x^2\widehat{Y}\},\\
2\hat{\gamma}\partial_x\hat{\gamma} &= 2\{\widehat{N} + e^{-\widehat{N}x}\widehat{Y}\}\partial_x\hat{\gamma} = 2\widehat{N}\partial_x\{e^{-\widehat{N}x}\widehat{Y}\} + 2e^{-\widehat{N}x}\widehat{Y}\widehat{P}\partial_x\hat{\gamma}\widehat{P}\\
&= 2\widehat{N}\partial_x\{e^{-\widehat{N}x}\widehat{Y}\} - e^{-\widehat{N}x}\widehat{Y}\hat{u}(x) = e^{-\widehat{N}x}\{-2N^2\widehat{Y} + 2\widehat{N}\partial_x\widehat{Y} - \widehat{Y}\hat{u}(x)\},
\end{aligned}$$

where

$$\hat{u}(x) = -2\partial_x \hat{P}\hat{\gamma}\hat{P} \in C^\infty(B(H_0)).$$

Putting these expressions into the first one of Eqs. (4.2.2), we arrive at the equalities

$$e^{-\hat{N}x}\{\partial_x^2\hat{Y} - \hat{Y}\hat{u}(x) - \hat{N}^2\hat{Y}\} = 0$$
$$\partial_x^2\{e^{-\hat{N}x}\hat{Y}\} - \{e^{-\hat{N}x}\hat{Y}\}\hat{u}(x) + 2\hat{N}\partial_x\{e^{-\hat{N}x}\hat{Y}\} = 0,$$

which are equivalent to the assertion of the lemma.

Similar assertions are valid for other linear equations of the left-hand side of Table II.

In what follows, we will consider solely for the operators $\hat{\Gamma}$ connected with the KdV equation, and, for the sake of simplicity, we will regard the space H_0 as one-dimensional. We will also assume that the conditions of Theorem 3.6.1 are met, which means the operators $\hat{\Gamma}$, $\hat{T}$ are bounded and convertible. Since the space H_0 is one-dimensional, the operators $\hat{a}(z)$, $\hat{a}$ in Eq. (4.0.1) are operators of multiplication by the function $a(z)$ and number a, respectively; these operators certainly depend on the variables x, t and are differentiable as many times as needed. Thus, in the case considered,

$$\hat{Y} = y(z)\hat{P}, \qquad \hat{u}(x) = u(x)\hat{P}, \qquad \hat{N} = iz.$$

To reduce the calculations, it is convenient to introduce the function

$$\psi_1(z) = \psi_1(z; x, t) = p(z)y(z) = p(z)\mathcal{E}(z)(\gamma(z) - iz)$$

instead of $y(z)$. This function, obviously, satisfies the equations

$$\text{(4.2.3)} \qquad \partial_x^2\psi_1(z) - \psi_1(z)u(x,t) + z^2\psi_1(z) = 0,$$
$$\text{(4.2.3}'\text{)} \quad \partial_x^2\{\mathcal{E}(-z)\psi_1(z)\} - \{\mathcal{E}(-z)\psi_1(z)\}u(x,t) + 2iz\{\mathcal{E}(-z)\psi_1(z)\} = 0$$

derived in Lemma 4.2.1. From results of the previous section it follows that the function $\psi_1(z)$ is found from the equation

$$\text{(4.2.4)} \quad \psi_1(z) + \rho_0(z)\{\nu(z)\psi_1(-z) + \int \frac{\mathcal{E}(-z)\mathcal{E}(-z')}{i(z'+z)}\psi_1(z')d\mu(z') + \mathcal{E}(-z)\} = 0,$$

where

$$\rho_0(z) = \varphi(z)p(z)r_1(z)\overline{r_1(-\bar{z})p(-\bar{z})},$$

while the potential $u(x,t)$ satisfies the KdV equation

$$4\partial_t u + 4\lambda\partial_x u + \partial_x^3 u - 3\partial_x(u^2) = 0.$$

Consider the function

$$\psi(z) = \psi(z; x, t) = \mathcal{E}(z)\left\{1 + \int \frac{\mathcal{E}(-z')\psi_1(z')}{i(z'-z)} d\mu(z')\right\} \tag{4.2.5}$$

defined by this equality for all $z \in C\backslash\Omega$, $-\infty < x, t < +\infty$. It is evident that in this domain it depends analytically on z, is infinitely differentiable with respect to x, t, and expands in the neighbourhood of the infinitely remote point z on the plane into the series

$$\psi(z) = \exp iz(x + (z^2 - \lambda)t)\{1 + i\sum_{k=1}^{\infty} z^{-k} v_k(x,t)\}, \tag{4.2.6}$$

where

$$v_k(x,t) = \int z^{k-1}\mathcal{E}(-z)\psi_1(z)d\mu(z),$$

whence, in particular it follows that

$$u(x,t) = -2\partial_x v_1(x,t).$$

LEMMA 4.2.2. *For all $z \in C\backslash\Omega$, the function $\psi(z; x, t)$ satisfies the Sturm–Liouville equation*

$$\partial_x^2\psi(z) - \psi(z)u(x,t) + z^2\psi(z) = 0$$

with the same potential $u(x,t)$ as in Lemma 4.2.1.

PROOF: Differentiating both parts of Eq. (4.2.5) twice with respect to x, we find that

$$\partial_x^2\psi(z) = -z^2\psi(z) + \mathcal{E}(z)\int \frac{2iz\partial_x\{\mathcal{E}(-z')\psi_1(z')\} + \partial_x^2\{\mathcal{E}(-z')\psi_1(z')\}}{i(z'-z)} d\mu(z')$$

and since, according to Eq. (4.2.3′),

$$\partial_x^2\{\mathcal{E}(-z')\psi_1(z')\} = \mathcal{E}(-z')\psi_1(z')u(x,t) - 2iz'\partial_x\{\mathcal{E}(-z')\psi_1(z')\},$$

then

$$\begin{aligned}\partial_x^2\psi(z) &= -z^2\psi(z) \\ &\quad + \mathcal{E}(z)\int \frac{2i(z-z')\partial_x\{\mathcal{E}(-z')\psi_1(z')\} + \mathcal{E}(-z')\psi_1(z')u(x,t)}{i(z'-z)} d\mu(z') \\ &= -z^2\psi(z) + \mathcal{E}(z)\{-2\partial_x \int \mathcal{E}(-z')\psi_1(z')d\mu(z')\} \\ &\quad + u(x,t)\{\psi(z) - \mathcal{E}(z)\} = (u(x,t) - z^2)\psi(z).\end{aligned}$$

Hence, $\partial_x^2\psi(z) - \psi(z)u(x,t) + z^2\psi(z) = 0$.

Analytic properties of the function $\psi(z) = \psi(z;x,t)$ enable us to use this solution to derive the formulas of expansion in eigenfunctions of the Sturm–Liouville operator. To derive the formulas, we shall employ expansion (4.2.6) which implies that for large $|z|$'s,

$$\psi(z;x,t)\psi(-z;y,t) = e^{iz(x-y)}\left\{1 + i\frac{v_1(x,t) - v_1(y,t)}{z} + \dots\right\},$$

$$\psi(-z;x,t)\psi(z;y,t) = e^{iz(y-x)}\left\{1 + i\frac{v_1(y,t) - v_1(x,t)}{z} + \dots\right\}.$$

Let us multiply both parts of these equalities by an arbitrary finite twice continuously differentiable function $f(y)$ and then integrate the first equality over the interval $(-\infty, x)$ and the second one over the interval (x, ∞). After integrating by parts, we obtain the equalities

$$\text{(4.2.7)}\qquad \begin{aligned} \psi(z;x,t)\int_{-\infty}^{x}\psi(-z;y,t)f(y)dy &= -\frac{f(x)}{iz} + \frac{\varepsilon_1(z)}{z^2},\\ \psi(-z;x,t)\int_{x}^{\infty}\psi(z;y,t)f(y)dy &= -\frac{f(x)}{iz} + \frac{\varepsilon_2(z)}{z^2}, \end{aligned}$$

in which functions $\varepsilon_1(z)$, $\varepsilon_2(z)$ are uniformly bounded in the bounded upper half-plane. Denote the loop, formed by a semi-circumference with the centre at the origin and the radius R lying in the upper plane and directed clockwise, by $C^+(R)$. Integrating equalities (4.2.7) along this loop, we find that

$$\text{(4.2.8)}\qquad \frac{1}{2\pi}\int_{C^+(R)}\{\psi(z;x,t)\int_{-\infty}^{x}\psi(-z;y,t)f(y)dy$$
$$+\psi(-z;x,t)\int_{x}^{\infty}\psi(z;y,t)f(y)dy\}dz = f(x) + O(R^{-1}).$$

We derive the expansion formulas from this equality by properly deforming the integration loop. For this, we need some auxilary propositions. As always, we shall denote by

$$\Omega_1^+, \Omega_2^+(\Omega_1^-, \Omega_2^-), \Omega_2^0$$

those parts of sets Ω_1, Ω_2 which lie in the upper (lower) half-plane and on the real axis. It is clear that the sets $\pi_1(\Omega_1^-) = \overline{\Omega_1^-}, \pi_1(\Omega_2^-) = \overline{\Omega_2^-}$ lie in the upper half-plane (with $\overline{\Omega_2^-} = \Omega_2^+$), while the set $\overline{\Omega_1^-}$ does not intersect the set $\Omega_1^+ \cup \Omega_2^+$. Henceforth, we assume that

$$\operatorname{dist}(\Omega_1^+, \bar\Omega_1^-) > 0, \qquad \operatorname{dist}(\Omega_2^+, \Omega_2^0) > 0, \qquad \operatorname{dist}(\Omega_1^+, \cup\, \bar\Omega_1^-, \Omega_2^+ \cup \Omega_2^0) > 0.$$

We could do without these constraints, but their introduction would allow to circumvent cumbersome limiting procedures. A typical support Ω possessing all these properties is shown in Fig. 3, where the set $\bar\Omega_1^-$ is given by a dashed line. Since the functions $\psi(z;x,t), \psi_1(z;x,t)$ are considered at a fixed value of t, this argument is omitted:

$$\psi(z;x,t) = \psi(z,x); \qquad \psi_1(z,x,t) = \psi_1(z,x).$$

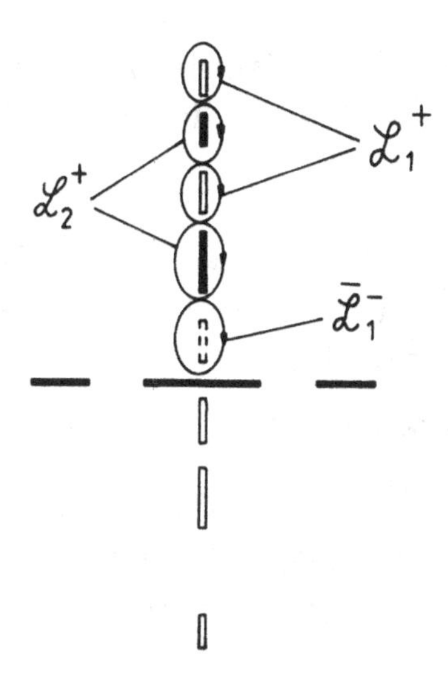

Figure 3
Left and right limit values of the function $\psi(z,x)$ on the real and imaginary axes are denoted by $\psi^{\pm}(z,x)$:

$\psi^{\pm}(z,x) = \psi(z \pm i0, x)(z \in (-\infty,\infty))$; $\quad \psi^{\pm}(z,x) = \psi(z \mp 0, x)(z \in (-i\infty, i\infty))$.

LEMMA 4.2.3. *The following equalities*

1) $\rho_0(z)\psi(-z,x) = -\psi_1(z,x), \quad z \in \Omega_1,$

2) $\rho_0(-z)\psi^{\pm}(z,x) = \rho_0(-z)(\pm\frac{1}{2} - \nu(-z))\psi_1(z,x) - \psi_1(-z,x), \quad z \in \Omega_2^0,$

3) $\rho_0(-z)\psi^{\pm}(z,x) = \rho_0(-z)(\pm\frac{1}{2i} - \nu(-z)\psi_1(z,x) - \psi_1(-z,x), \quad z \in \Omega_2^+ \cup \Omega_2^-$

are valid on the support $\Omega = \Omega_1 \cup \Omega_2$ *of the measure* $d\mu(z)$.

PROOF: When $z \in \Omega_1$, the integral in Eq. (4.2.4) has no singularities and $\nu(z) = 0$. So, at such values of z,

$$0 = \psi_1(z) + \rho_0(z)\mathcal{E}(-z)\{1 + \int \frac{\mathcal{E}(-z')\psi_1(z')}{i(z'+z)} d\mu(z')\}$$
$$= \psi_1(z) + \rho_0(z)\psi(-z),$$

whence the first of the equalities to be proved follows. Further, since the measure $d\mu(z)$ on the set Ω_2 coincides with the Lebesgue measure $(2\pi)^{-1}|dz|$, we have, according to the Plemelj–Sochocki formulas,

$$\psi(z \pm i0, x) = \mathcal{E}(z)\left\{1 + \int \frac{\mathcal{E}(-z')\psi_1(z')}{i(z'-z)} d\mu(z') \pm \frac{1}{2}\mathcal{E}(-z)\psi_1(z,x)\right\},$$

$$\psi(z \mp 0, x) = \mathcal{E}(z)\left\{1 + \int \frac{\mathcal{E}(-z')\psi_1(z')}{i(z'-z)} d\mu(z') \pm \frac{1}{2i}\mathcal{E}(-z)\psi_1(z,x)\right\}.$$

Relations 2 and 3 follow immediately from these equalities and Eq. (4.2.4).

Note that the function $\rho_0(z)$ differs from zero almost everywhere in the measure $d\mu(z)$. This permits us to rewrite formulas 1,2,3 in the form

$$\psi(-z,x) = -b(z)\psi_1(z,x) \qquad (z \in \Omega_1), \tag{4.2.9}$$

$$\psi^{\pm}(z,x) = a^{\pm}(z)\psi_1(z,x) - b(-z)\psi_1(-z,x) \qquad (z \in \Omega_2), \tag{4.2.10}$$

where

$$a^{\pm}(z) = \begin{cases} \pm\frac{1}{2} - \nu(-z) & z \in \Omega_2^0 \\ \pm\frac{1}{2i} - \nu(-z) & z \in \Omega_2^+ \cup \Omega_2^-, \end{cases} \tag{4.2.11}$$
$$b(z) = \rho_0(z)^{-1}.$$

In particular, it follows from this that

$$\begin{aligned}\{\psi(z,x)\psi(-z,y)\}^{\pm} &= a^{\pm}(z)a^{\mp}(-z)\psi_1(z,x)\psi_1(-z,y) \\ &- a^{\mp}(-z)b(-z)\psi_1(-z,x)\psi_1(-z,y) \\ - a^{\pm}(z)b(z)\psi_1(z,x)\psi_1(z,y) &+ b(z)b(-z)\psi_1(-z,x)\psi_1(-z,y), (z \in \Omega_2).\end{aligned} \tag{4.2.10'}$$

LEMMA 4.2.4. *Let $\mathcal{L}_1^+$, $\bar{\mathcal{L}}_1^-$, $\mathcal{L}_2^+$ be loops oriented clockwise surrounding the sets Ω_1^+, $\bar{\Omega}_1^-$, Ω_2^+, respectively, and containing no other points of the support of the measure $d\mu(z)$ (see Fig. 3). Then*

$$\frac{1}{2\pi}\int_{\mathcal{L}_1^+} \psi(z,x)\psi(-z,y)dz = -\int_{\Omega_1^+} \psi(-z,x)\psi(-z,y)\rho_0(z)d\mu(z),$$

$$\frac{1}{2\pi}\int_{\bar{\mathcal{L}}_1^-} \psi(z,x)\psi(-z,y)dz = \int_{\Omega_1^-} \psi(-z,x)\psi(-z,y)\rho_0(z)d\mu(z),$$

$$\frac{1}{2\pi}\int_{\mathcal{L}_2^+} \psi(z,x)\psi(-z,y)dz = \int_{\Omega_2^+} \{(\nu(-z)-\nu(z))\psi_1(z,x)\psi_1(-z,y)$$
$$-b(z)\psi_1(z,x)\psi_1(z,y) + b(-z)\psi_1(-z,x)\psi_1(-z,y)\}d\mu(z).$$

PROOF: Denote the domains inside the loops $\mathcal{L}_1^+$, $\bar{\mathcal{L}}_1^-$ by D_1^+, D_1^-. By definition, the function $\psi(z,x)$ is holomorphic in the domain D_1^-, and the function $\psi(-z,y)$ is holomorphic in the domain D_1^+. Besides,

$$\psi(z,x) = \mathcal{E}(z,x)\int_{\Omega_1^+} \frac{\mathcal{E}(-z',x)\psi_1(z',x)}{i(z'-z)}d\mu(z') + \alpha^+(z), \qquad z \in D_1^+,$$

$$\psi(-z,y) = \mathcal{E}(-z,y)\int_{\Omega_1^-} \frac{\mathcal{E}(-z',y)\psi_1(z',y)}{i(z'+z)}d\mu(z') + \alpha^-(z), \qquad z \in D_1^-,$$

where the functions

$$\alpha^{+}(z) = \mathcal{E}(z,x)\left\{1 + \int\limits_{\Omega\setminus\Omega_1^+} \frac{\mathcal{E}(-z',x)\psi_1(z',x)}{i(z'-z)}\,d\mu(z')\right\},$$

$$\alpha^{-}(z) = \mathcal{E}(-z,y)\left\{1 + \int\limits_{\Omega\setminus\Omega_1^-} \frac{\mathcal{E}(-z',y)\psi_1(z',y)}{i(z'+z)}\,d\mu(z')\right\},$$

are holomorphic in the domains D_1^+ and D_1^-, respectively. Therefore,

$$\begin{aligned}
&\frac{1}{2\pi}\int\limits_{\mathcal{L}_1^+} \psi(z,x)\psi(-z,y)\,dz \\
&\qquad = \frac{1}{2\pi}\int\limits_{\mathcal{L}_1^+} \Big(\psi(-z,y)\mathcal{E}(z,x)\int\limits_{\Omega_1^+} \frac{\mathcal{E}(-z',x)\psi_1(z',x)}{i(z'-z)}\,d\mu(z')\Big)\,dz,\\
&\frac{1}{2\pi}\int\limits_{\overline{\mathcal{L}_1^-}} \psi(z,x)\psi(-z,y)\,dz \\
&\qquad = \frac{1}{2\pi}\int\limits_{\overline{\mathcal{L}_1^-}} \Big(\psi(z,x)\mathcal{E}(-z,y)\int\limits_{\Omega_1^-} \frac{\mathcal{E}(-z',y)\psi_1(z',y)}{i(z'+z)}\,d\mu(z')\Big)\,dz.
\end{aligned}$$

Changing the order of integration in the right-hand side of these equalities and then employing the Cauchy formula and formula 1 of Lemma 4.2.3, we find that

$$\begin{aligned}
\frac{1}{2\pi}\int\limits_{\mathcal{L}_1^+} \psi(z,x)\psi(-z,y)\,dz &= \int\limits_{\Omega_1^+} \mathcal{E}(z',x)\mathcal{E}(-z',x)\psi(-z',y)\psi_1(z',x)\,d\mu(z')\\
&= -\int\limits_{\Omega_1^+} \psi(-z',x)\psi(-z',y)\rho_0(z')\,d\mu(z')\\
&= -\int\limits_{\Omega_1^+} \psi_1(z',x)\psi_1(z',y)b(z')\,d\mu(z'),\\
\frac{1}{2\pi}\int\limits_{\overline{\mathcal{L}_1^-}} \psi(z,x)\psi(-z,y)\,dz &= -\int\limits_{\Omega_1^-} \mathcal{E}(z',y)\mathcal{E}(-z',y)\psi(-z',x)\psi_1(z',y)\,d\mu(z')\\
&= \int\limits_{\Omega_1^-} \psi(-z',x)\psi(-z',y)\rho_0(z')\,d\mu(z')\\
&= \int\limits_{\Omega_1^-} \psi_1(z',x)\psi_1(z',y)b(z')\,d\mu(z').
\end{aligned}$$

Contracting the loop $\mathcal{L}_2^+$ towards segments from the set Ω_2^+ passed twice and noticing that, by condition, $d\mu(z) = (2\pi)^{-1}|dz| = (2\pi i)^{-1}dz$ on this set, we get the following equality,

$$\frac{1}{2\pi}\int_{\mathcal{L}_2^+} \psi(z,x)\psi(-z,y)dz = i\int_{\Omega_2^+}\{(\psi(z,x)\psi(-z,y))^+ - (\psi(z,x)\psi(-z,y))^-\}d\mu(z).$$

To calculate the integrand on the right-hand side of this expression, one may use formulas (4.2.10′) and (4.2.11). After elementary calculation, we obtain

$$\begin{aligned}(\psi(z,x)\psi(-z,y))^+ - (\psi(z,x)\psi(-z,y))^- &= -i(\nu(-z)-\nu(z))\psi_1(z,x)\psi_1(-z,y)\\ &\quad + ib(z)\psi_1(z,x)\psi_1(z,y) - ib(-z)\psi_1(-z,x)\psi_1(-z,y),\end{aligned}$$

and

$$\begin{aligned}\frac{1}{2\pi}\int_{\mathcal{L}_2^+}\psi(z,x)\psi(-z,y)dz &= \int_{\Omega_2^+}\{(\nu(-z)-\nu(z))\psi_1(z,x)\psi_1(-z,y)\\ &\quad - b(z)\psi_1(z,x)\psi_1(z,y) + b(-z)\psi_1(-z,x)\psi_1(-z,y)\}d\mu(z).\end{aligned}$$

The conditions that (3.6.1) must be real impose on the functions $\rho_0(z)$, $\nu(z)$ the constraints $\rho_0(-\bar{z}) = \overline{\rho_0(z)}$, $\nu(\bar{z}) = \overline{\nu(z)}$. Therefore,

$$\begin{aligned}&\rho_0(z) = \varphi(z)|\rho_0(z)|(z\in\Omega_1\cup\Omega_2^+\cup\Omega_2^-), \quad \rho_0(z) = \overline{\rho(-z)} \quad (z\in\Omega_2^0),\\ &\overline{\nu(z)} = \nu(-z)(z\in\Omega_2^+\cup\Omega_2^-), \qquad \nu(z) = \overline{\nu(z)} \quad (z\in\Omega_2^0).\end{aligned} \tag{4.2.12}$$

We assume henceforth that the function $\nu(z)$ is chosen to be even, i.e., real-valued:

$$\nu(z) = \nu(-z) = \overline{\nu(z)}, \qquad z\in\Omega_2^0. \tag{4.2.13}$$

As a consequence of these constraints, the functions $\psi_1(z,x)$, $\psi(z,x)$ possess the following properties:

$$\overline{\psi_1(-\bar{z},x)} = \psi_1(z,x)(z\in\Omega); \qquad \overline{\psi(-\bar{z},x)} = \psi(z,x)(z\in C\backslash\Omega).$$

Indeed, since

$$d\mu(z) = d\mu(-\bar{z}), \quad \overline{\rho(-\bar{z})} = \rho(z), \quad \overline{\nu(-\bar{z})} = \overline{\nu(\bar{z})} = \nu(z), \quad \overline{\mathcal{E}(-\bar{z})} = \mathcal{E}(z),$$

we have, by carrying out the complex conjugation in Eq. (4.2.4) and by the substitution, $z' \to -\bar{z}'$, $z \to -\bar{z}$, the functions $\psi_1(z,x)$, $\overline{\psi_1(-\bar{z},x)}$ that satisfy the same equation. But this equation has a unique solution, hence $\psi_1(z,x) = \overline{\psi_1(-\bar{z},x)}$. Employing this equality and carrying out similar operations in both parts of Eq. (4.2.5), we find that $\psi(z,x) = \overline{\psi(-\bar{z},x)}$.

Note that the relations obtained in Lemma 4.2.4 assume such a form:

$$
\begin{aligned}
\frac{1}{2\pi}\int_{\mathcal{L}_1^+} \psi(z,x)\psi(-z,y)\,dz &= \int_{\Omega_1^+} \psi_1(z,x)\psi_1(z,y)|b(z)|\,d\mu(z),\\
\frac{1}{2\pi}\int_{\overline{\mathcal{L}_1^-}} \psi(z,x)\psi(-z,y)\,dz &= \int_{\Omega_1^-} \psi_1(z,x)\psi_1(z,y)|b(z)|\,d\mu(z),\\
\frac{1}{2\pi}\int_{\mathcal{L}_2^+} \psi(z,x)\psi(-z,y)\,dz &= \int_{\Omega_2^+} \{\psi_1(z,x)\psi_1(z,y)|b(z)|\\
&\quad + \psi_1(-z,x)\psi_1(-z,y)|b(-z)|\}\,d\mu(z),
\end{aligned} \tag{4.2.14}
$$

while the functions $\psi_1(z,x)$, $\psi_1(-z,x)$ are real on the sets over which the integration is carried out on the right-hand side of these equalities. Let us now proceed to the derivation of the decomposition formula. Since both functions $\psi(z,x)$, $\psi(-z,y)$ are holomorphic in the upper half-plane outside the set $\Omega_1^+ \cup \overline{\Omega_1^-} \cup \Omega_2^+$, the integral in the left-hand side of Eq. (4.2.8) is equal to the sum of integrals taken along the loops $\mathcal{L}_1^+$, $\overline{\mathcal{L}_2^-}$, $\mathcal{L}_2^+$ and also along the loop $\mathcal{L}_0^+$ consisting of the upper edge of the segment $(-R,R)$. Denote these integrals by J_1^+, J_1^-, J_2^+, J_0^+ and rewrite (4.2.8) in an equivalent form,

$$
J_1^+ + J_1^- + J_2^+ + J_0^+ = f(x) + O(R^{-1}). \tag{4.2.15}
$$

To calculate the first three integrals, we shall use formulas (4.2.14) whose right-hand sides are symmetric with respect to the arguments x, y. To give an example, we are going to calculate the integral J_1^+. From the equalities

$$
\begin{aligned}
\frac{1}{2\pi}\int_{\mathcal{L}_1^+} \psi(z,x)\psi(-z,y)\,dz &= \int_{\Omega_1^+} \psi_1(z,x)\psi_1(z,y)|b(z)|\,d\mu(z),\\
\frac{1}{2\pi}\int_{\mathcal{L}_1^+} \psi(z,y)\psi(-z,x)\,dz &= \int_{\Omega_1^+} \psi_1(z,y)\psi_1(z,x)|b(z)|\,d\mu(z),
\end{aligned}
$$

it follows that

$$
\begin{aligned}
\frac{1}{2\pi}\int_{\mathcal{L}_1^+} \{\psi(z,x)\int_{-\infty}^{x} \psi(-z,y)f(y)\,dy\}\,dz &= \int_{\Omega_1^+} (\psi_1(z,x)\int_{-\infty}^{x} \psi_1(z,y)f(y)\,dy)|b(z)|\,d\mu(z),\\
\frac{1}{2\pi}\int_{\mathcal{L}_1^+} \{\psi(-z,x)\int_{x}^{\infty} \psi(z,y)f(y)\,dy\}\,dz &= \int_{\Omega_1^+} (\psi_1(z,x)\int_{x}^{\infty} \psi_1(z,y)f(y)\,dy)|b(z)|\,d\mu(z),
\end{aligned}
$$

and, therefore,

$$J_1^+ = \int_{\Omega_1^+} \psi_1(z,x)\{\int_{-\infty}^{x} \psi_1(z,y)f(y)dy + \int_{x}^{\infty} \psi_1(z,y)f(y)dy\}|b(z)|d\mu(z)$$

$$= \int_{\Omega_1^+} (\psi_1(z,x)\{\int_{-\infty}^{\infty} \psi_1(z,y)f(y)dy)|b(z)|d\mu(z)$$

The other two integrals are calculated in the same manner. Introducing for brevity, the notation

$$\psi_1(z,f) = \int_{-\infty}^{\infty} \psi_1(z,y)f(y)dy, \qquad \psi(z,f) = \int_{-\infty}^{\infty} \psi(z,y)f(y)dy,$$

we obtain the equality

$$J_1^+ + \bar{J}_1^- + J_2^+ = \int_{\Omega_1^+ \cup \bar{\Omega}_1^-} \psi_1(z,x)\psi_1(z,f)|b(z)|d\mu(z)$$

$$+ \int_{\Omega_2^+} \{\psi_1(z,x)\psi_1(z,f)|b(z)| + \psi_1(-z,x)\psi_1(-z,f)|b(-z)|\}d\mu(z).$$

Now, we have to calculate the remaining integral J_0^+. Let the number R be so large that the set Ω_2^0 lies entirely inside the segment $(-R,R)$. The complement of the set Ω_2^0 to the segment $(-R,R)$ will be denoted by $\Omega_2(R)$, while the parts of these sets lying on the positive semi-axis will be denoted by $\Omega_2^0(+)$, $\Omega_2(R^+)$, respectively.

Since the sets Ω_2^0, $\Omega_2(R)$ are symmetric with respect to the origin, the integrals of the function $F(z)$ over these sets equal the integrals of the function $F(z)+F(-z)$ taken over positive parts of these sets.

Further, since the set $\Omega_2(R)$ belongs to the domain where the function $\psi(z,x)\psi(-z,y)$ is holomorphic, then

$$(\psi(z,x)\psi(-z,y))^+ = \psi(z,x)\psi(-z,y)$$

and

$$\text{(4.2.16)} \quad \frac{1}{2\pi}\int_{\Omega_2(R)} (\psi(z,x)\psi(-z,y))^+ dz = \frac{1}{2\pi}\int_{\Omega_2(R)} \psi(z,x)\psi(-z,y)dz$$

$$= \frac{1}{2\pi}\int_{\Omega_2(R^+)} \{\psi(z,x)\psi(-z,y) + \psi(-z,x)\psi(z,y)\}dz.$$

The fact that the function $\nu(z)$ is even, together with Eq. (4.2.11), implies that

$$a^+(z)a^-(-z) = -\frac{1}{4} + \nu(-z)\nu(z) - \frac{1}{2}(\nu(z) - \nu(-z)) = \nu(z)^2 - \frac{1}{4},$$

$$a^+(z) + a^-(-z) = -2\nu(z) = -2\nu(-z).$$

Combining these equalities and Eq. (4.2.10), we find that on Ω_2^0 the doubled even part of the function $(\psi(z,x)\psi(-z,y))^+$ is equal to

$$\alpha(z)\{\psi_1(-z,x)\psi_1(z,y)+\psi_1(z,x)\psi_1(-z,y)\} \\ +2\nu(z)\{b(z)\psi_1(z,x)\psi_1(z,y)+b(-z)\psi_1(-z,x)\psi_1(-z,y)\},$$

where

$$\alpha(z)=\nu(-z)\nu(z)+b(z)b(-z)-\frac{1}{4}=\nu(z)^2+|b(z)|^2-\frac{1}{4}.$$

That is why

$$\frac{1}{2\pi}\int_{\Omega_2^0}(\psi(z,x)\psi(-z,y))^+dz=\frac{1}{2\pi}\int_{\Omega_2^0(+)}\{\alpha(z)(\psi_1(-z,x)\psi_1(z,y) \\ +\psi_1(z,x)\psi_1(-z,y))+2\nu(z)(b(z)\psi_1(z,x)\psi_1(z,y) \\ +b(-z)\psi_1(-z,x)\psi_1(-z,y))\}dz.$$

The right-hand side of Eqs. (4.2.16) and (4.2.17) are symmetric with respect to the arguments x,y. Using the symmetry and proceeding in the same manner as in the calculation of the integral J_1^+, we obtain the following expression for the integral J_0^+:

$$J_0^+=\int_{\Omega_2(R^+)}\{\psi(z,x)\psi(-z,f)+\psi(-z,x)\psi(z,f)\}d\mu(z) \\ +\int_{\Omega_2^0(+)}\{\alpha(z)(\psi_1(z,x)\psi_1(-z,f)+\psi_1(-z,x)\psi_1(z,f)) \\ +2\nu(z)(b(z)\psi_1(z,x)\psi_1(z,f)+b(-z)\psi_1(-z,x)\psi_1(-z,f))\}d\mu(z).$$

(We have taken into account that on the real axis $d\mu(z)=(2\pi)^{-1}dz$). Substituting into Eq. (4.2.15) the expressions found for the integrals J_1^+, $\bar{J}_1^-$, J_2^+, J_0^+ and sending R to infinity, we arrive at the following decomposition of the function $f(x)$:

$$f(x)=\int_{\Omega_1^+\cup\Omega_1^-}\psi_1(z,x)\psi_1(z,f)|b(z)|d\mu(z) \\ +\int_{\Omega_2^+}\{\psi_1(z,x)\psi_1(z,f)|b(z)|+\psi_1(-z,x)\psi_1(-z,f)|b(-z)|\}d\mu(z) \\ +\int_{\Omega_2^0(+)}\{\alpha(z)(\psi_1(z,x)\psi_1(-z,f)+\psi_1(-z,x)\psi_1(z,f)) \\ +2\nu(z)(\psi_1(z,x)\psi_1(z,f)b(z)+\psi_1(-z,x)\psi_1(-z,f)b(-z))\}d\mu(z) \\ +\int_{\Omega_2(\infty+)}\{\psi(z,x)\psi(-z,f)+\psi(-z,x)\psi(z,f)\}d\mu(z).$$

A more compact and graphic form can be given to this expression. To this end, we will introduce solutions $\omega(z,x)$ of the Sturm–Liouville equation (4.2.3) which are defined at real values of the spectral parameter z^2 by the following formulas:

$$\omega(z,x) = \begin{cases} \overline{b(z)\psi_1(z,x)} \equiv b(-\bar{z})\psi_1(-\bar{z},x) & z \in \Omega \\ \overline{\psi(z,x)} \equiv \psi(-\bar{z},x) & z \notin \Omega. \end{cases}$$

Let us denote

$$\omega(z,f) = \int_{-\infty}^{\infty} \omega(z,y)f(y)dy.$$

Using this denotation and taking into account the equality $d\mu(z) = d\mu(-\bar{z})$ and the evenness of the function $\nu(z)$, we can rewrite the above decomposition of the function $f(x)$ in the form

$$f(x) = \int_{\Omega\cup(R^1\setminus\Omega_2^0)} \overline{\omega(z,x)}\omega(z,f)n(z)d\mu(z)$$
$$+2\int_{\Omega_2^0} \overline{\omega(z,x)}\omega(-z,f)\nu(z)\rho_0(-z)d\mu(z),$$

where

$$n(z) = \begin{cases} |\rho_0(z)| & z \in \Omega_1 \cup \Omega_2^+ \cup \Omega_2^- \\ 1 + |\rho_0(z)|^2\left(\nu(z)^2 - \frac{1}{4}\right) & z \in \Omega_2^0 \\ 1 & z \in R^1\setminus\Omega_2^0 \end{cases} \tag{4.2.18'}$$

The form of the decomposition obtained implies that the Sturm–Liouville operator (4.2.3) with the potential $u(x,t)$ has the following spectral properties:

1) The negative spectrum consists of two parts:

$$S_1^- = \{\lambda : \lambda = z^2,\ z \in \Omega_1\}, \qquad S_2^- = \{\lambda : \lambda = z^2, z \in \Omega_2\}.$$

The spectrum S_2^- is double and, of course, absolutely continuous. The spectrum S_1^- is single and its character is determined by properties of the measure $d\mu(z)$ on the set Ω_1. Since the measure may be defined on this set in quite an arbitrary fashion, this part of the spectrum can contain absolutely continuous portions, as well as singular and discrete parts.

2) On the positive semi-axis, the spectrum is double and absolutely continuous. The set where the equalities $\nu(z) = 0$, $|\rho_0(z)|^2 = 4$ are valid does not belong to the spectrum, forming gaps in it. Note that in this case, one may not use Theorem 3.6.1 directly, because inequality 5 of the theorem becomes an equality. Thus, when $\nu(z) = 0$ $|\rho_0(z)|^2 = 4$, one has to use the generalization of this theorem which was mentioned above.

These properties of Sturm–Liouville operators give an indirect characteristic of the obtained solutions to the KdV equation. In the general case, their behavior at $x \to \pm\infty$ can be rather capricious and can hardly be described in exact terms. But a certain understanding of this class is not impossible.

For instance, if the measure support lies solely on the imaginary axis and $\nu(z) = 0$, then the obtained solutions $u(x,s)$ meet the condition

$$-2\kappa^2 \leqslant u(x,t) \leqslant 0,$$

where $-\kappa^2 = \inf_{z\in\Omega} z^2$. The solutions cannot tend to zero when $x \to +\infty(x \to -\infty)$, if the set $\Omega^-(\Omega^+)$ has nonzero limit points. It can be shown that such potentials are limits of nonreflective ones. In particular, finite-zone solutions have this property. In addition, the set Ω_2 is empty and the measure on each of the sets Ω_1^+, Ω_1^- is singular, then the solution $u(x,t)$ at $x \to \pm\infty$ cannot pass into a quasi-periodic mode.

Note, finally, that in certain simple cases it is sometimes possible to separate leading terms of the asyptotics of the solution $u(x,t)$ when $x \to \pm\infty$.

§3 The KP Equation

The operators $\widehat{\Gamma}$ associated with this equation are given by formula (3.2.II(1)) while the auxilary condition is equivalent to equation (3.3.II(1)). If in the capacity of H_0 we take a one-dimensional space, then operators $\widehat{P}(\hat{\gamma} - \widehat{N})\widehat{P}$ $(\hat{\gamma} = \widehat{\Gamma}^{-1}\partial_x\widehat{\Gamma})$ will assume the form $\widehat{P}(\hat{\gamma} - \widehat{N})\widehat{P} = u(x,t,y)\widehat{P}$, where $u = u(x,t,y)$ is a scalar infinitely differential function (Eq. (4.0.1)). In this case, $[\partial_y\widehat{P}(\hat{\gamma} - \widehat{N})\widehat{P}, \partial_x\widehat{P}(\hat{\gamma} - \widehat{N})\widehat{P}] = 0$, and Eq. (3.2.II(1)), which is satisfied by the operators $\widehat{P}(\hat{\gamma} - \widehat{N})\widehat{P}$, is reduced to the KP equation

$$\partial_x(4\partial_t u + \partial_x^3 u + 6(\partial_x u)^2) + 3\varepsilon^2\partial_y^2 u = 0 \tag{4.3.1}$$

in the function $u = u(x,t,y)$. Let us take the space $L^2_\mu(\Omega)$ of the scalar functions $f(z)$ in the capacity of the extended space H and define the projection operator $\widehat{P}$ by the equality

$$\widehat{P}(f,(z)) = \int_\Omega f(z)p(z)d\mu(z),$$

$$p(z) \in L^2_\mu(\Omega), \qquad \int_\Omega p(z)d\mu(z) = 1;$$

we also choose the operators $\hat{a}(\xi)$, $\widehat{N}$ in formulas (3.2.II(1)), (3.3.II(1)) to be equal to the operator of multiplication by the function $\alpha(z) : \hat{a}(\xi) = \widehat{N} = \alpha(z)$. If the support Ω of the measure $d\mu(z)$ lies in the domain where this function is one-sheeted, then the formal solution of Eq. (3.3.II(1)) in operators $\widehat{C}_2(\xi)$ has the form

$$\widehat{C}_2(\xi)(f(z)) = m(z,\xi)f(z) + r(z,\xi)\fint_\Omega \frac{f(z')p(z')}{\alpha(z') - \alpha(z)}d\mu(z'),$$

and the operator $\widehat{\Gamma}$ has the form

$$\widehat{\Gamma}(f(z)) = \int_{\Sigma} \widehat{C}_1(\xi)\{e^{\theta(z)}(m(z,\xi)f(z) + r(z,\xi)\int_{\Omega} \frac{f(z')p(z')}{\alpha(z')-\alpha(z)} d\mu(z'))\} d\nu(\xi),$$

where

$$\theta(z) = \alpha(z)x - \alpha(z)^3 t - \varepsilon^{-1}\alpha(z)^2 y.$$

Besides,

$$\partial_x \widehat{\Gamma} = \widehat{\Gamma}\alpha(z) - \int_{\Sigma} \widehat{C}_1(\xi)(e^{\theta(z)} r(z,\xi))\widehat{P} d\nu(\xi),$$

$$(\widehat{\gamma} - \alpha(z)\widehat{P} = (\gamma(z) - \alpha(z))\widehat{P} = g(z)\widehat{P}, \qquad u = \int_{\Omega} g(z)p(z)d\mu(z),$$

and the function $g(z)$ is found from the equation

$$\int_{\Sigma} \widehat{C}_1(\xi)(e^{\theta(z)} m(z,\xi)g(z)d\nu(\xi)$$

$$= -\int_{\Sigma} \widehat{C}_1(\xi)\left\{e^{\theta(z)} r(z,\xi)(1 + \int_{\Omega} \frac{g(z')p(z')}{\alpha(z')-\alpha(z)} d\mu(z'))\right\} d\nu(\xi).$$

We confine ourselves to considering a particular case of this equation when the measure $d\nu(\xi)$ is concentrated in two points ξ_1, ξ_2, a unit mass in each. If the parameters $\widehat{C}_1(\xi)$, $m(z,\xi)$, $r(z,\xi)$ satisfy the conditions:

$$\widehat{C}_1(\xi_1) = I, \qquad r(z,\xi_1) = r(z), \qquad r(z,\xi_2) = 0,$$
$$\widehat{C}_1(\xi_2) = \widehat{C}, \qquad m(z,\xi_2) = m(z), \qquad m(z,\xi_1) = 0,$$

then this equation assumes the form

$$\widehat{C}(e^{\theta(z)} m(z)g(z)) = -e^{\theta(z)} r(z)\{1 + \int_{\Omega} \frac{g(z')p(z')}{\alpha(z')-\alpha(z)} d\mu(z')\}.$$

Let the measure $d\mu(z)$ be concentrated on a piecewise-smooth curve $\mathcal{L}$ surrounding a domain D, $z_0 \in D$,

$$p(z)d\mu(z) = (2\pi i)^{-1}(z - z_0)^{-1} dz, \qquad \alpha(z) = (z - z_0)^{-1}$$

and

$$\widehat{C}(f,(z)) = \widehat{\pi}(f(z)) = f(\pi(z)),$$

where $\pi(z)$ is a one-to-one mapping of the curve $\mathcal{L}$ onto itself, satisfying the condition $\pi^2 = I$. In this case, the function $g(z)$ is found from the equation

$$\widehat{\pi}(e^{\theta(z)} m(z)g(z)) = -e^{\theta(z)} r(z)\{1 - \frac{(z-z_0)}{2\pi i} \int_{\mathcal{L}} \frac{g(z')}{z'-z} dz'\}. \tag{4.3.2}$$

Consider a function

$$\psi(z) = 1 - \frac{z - z_0}{2\pi i} \int_{\mathcal{L}} \frac{g(z')}{z' - z} dz'$$

holomorphic in the domain D. In the neighbourhood of the point z_0, it expands into the Taylor series:

$$\psi(z) = 1 + \sum_{k=1}^{\infty} (z - z_0)^k v_k \qquad v_k = v_k(x, t, y)),$$

where

$$v_k = \frac{1}{2\pi i} \int_{\mathcal{L}} (z - z_0)^{-k} g(z) dz,$$

whence, in particular, it follows that the function

$$v_1 = \frac{1}{2\pi i} \int_{\mathcal{L}} (z - z_0)^{-1} g(z) dz = \int_{\mathcal{L}} g(z) p(z) d\mu(z) = u(x, t, y)$$

is a solution to the KP equation. According to the Plemelj–Sochocki formula, inner limit values $\psi^+(z)$ of the function $\psi(z)$ on the loop $\mathcal{L}$ are equal to

$$\psi^+(z) = 1 - \frac{z - z_0}{2\pi i} \not\int \frac{g(z')}{z' - z} dz' - \frac{z - z_0}{2} g(z).$$

Combining this formula and Eq. (4.3.2), we have

$$e^{\theta(z)} r(z) \psi^+(z) = -\hat{\pi}(e^{\theta(z)} m(z) g(z)) + e^{\theta(z)} \frac{z - z_0}{2} r(z) g(z),$$

and if

$$m(z) = -\frac{1}{2}(z - z_0) r(z),$$

then

$$e^{\theta(z)} r(z) \psi^+(z) = \hat{\pi}(g_1(z)) + g_1(z),$$

where $g_1(z) = \frac{1}{2}(z - z_0) r(z) e^{\theta(z)} g(z)$. Further, since $\pi^2 = I$,

$$\hat{\pi}(e^{\theta(z)} r(z) \psi^+(z)) = \hat{\pi}^2(g_1(z)) + \hat{\pi}(g_1(z)) = g_1(z) + \hat{\pi}(g_1(z)) = e^{\theta(z)} r(z) \psi^+(z).$$

Therefore, the function $\psi(z)$ possesses the following properties:

1) it is holomorphic in the domain D, and $\psi(z_0) = 1$;

2) its limit values on the boundary $\mathcal{L}$ of this domain are related by the expression

$$e^{\theta(\pi(z))} r(\pi(z)) \psi^+(\pi(z)) = e^{\theta(z)} r(z) \psi^+(z). \tag{4.3.3}$$

The parameters π, $r(z)$ can be chosen in such a manner that the function $\psi_1(z) = e^{\theta(z)} r(z) \psi(z)$ becomes a generalized Baker–Akhiezer function introduced by Krichever. In this case, the solutions obtained coincide with those found by the Krichever method [14].

Indeed, a compact Riemann surface F for the kind g can be realized as a fundamental polygon M from the corresponding group of linear-fractional transformations of the unit circle. The number of sides of the polygon equals $4g$ and its sides

$$a_1,\ b_1,\ a_1^{-1},\ b_1^{-1},\dots,a_g,\ b_g,\ a_g^{-1},\ b_g^{-1}$$

are circumference arcs and form a boundary $\mathcal{L}_0$ of the domain M. Denote by $\alpha_i,\ \beta_i(i=1,2,\dots,g)$ generators of this group that transform the sides a_i into a_i^{-1} and b_i into b_i^{-1} : $\alpha_1(a_i)=a_i^{-1},\ \beta_i(b_i)=b_i^{-1}$. Define a one-to-one correspondence π_0 of the boundary $\mathcal{L}_0$ onto itself:

$$\pi_0(z)=\begin{cases}\alpha_i^s(z), & z\in a_i^s\\ \beta_i^s(z), & z\in b_i^s\end{cases}\qquad (s=1,\ -1)$$

Let in Eq. (4.3.2) $D=M,\ \mathcal{L}=\mathcal{L}_0,\ \pi=\pi_0$ and $r(z)=p(z_0)p(z)^{-1}$, where

$$p(z)=\prod_{i=1}^{g}(z-z_i)\qquad (z_i\in D,\ z_i\neq z_0).$$

Then the function

$$\psi_1(z)=e^{\theta(z)}r(z)\psi(z)=e^{\theta(z)}r(z)\left\{1-\frac{z-z_0}{2\pi i}\int_{\mathcal{L}_0}\frac{g(z')}{z'-z}dz'\right\}$$

is meromorphic in the domain $M\backslash z_0$, and has simple poles at points $z_1,z_2,\dots,z_g$, while in the neighbourhood of the point z_0

$$\psi_1(z)=e^{\theta(z)}\{1+\sum_{k=1}^{\infty}(z-z_0)^k\tilde{v}_k\},\tag{4.3.4}$$

where the function $\tilde{v}_1=\tilde{v}_1(x,t,y)$ differs from $u(x,t,y)$ by a constant term which means that $\tilde{v}_1$ also satisfies the KP equation (4.3.1).

Besides, according to Eq. (4.3.3), $\psi_1^+(\pi_0(z))=\psi_1^+(z)$, that is, the function $\psi_1(z)$ has the same limit values in congruent points of the boundary $\mathcal{L}_0$. That is why the function $\psi_1(z)$ can be analytically extended to the unit circle, as an automorphic function having essential singularities at points congruent to z_0 and simple poles at points congruent to $z_1,\ z_2,\dots,z_g$. By identifying the congruent sides a_i,a_i^{-1} and b_i,b_i^{-1}, we obtain the Riemann surface F on which, according to the above reasoning, the function $\psi_1(z)$ is one-valued, possesses the only essential singularity at the point z_0 (in whose neighbourhood it is representable in the form (4.3.4)) and simple poles at the points $z_1,\ z_2,\dots,z_g$. Thus $\psi_1(z)$ is a generalized Baker–Akhiezer function on this Riemann surface. The nonspecific character of the divisor $z_1,z_2,\dots,z_g$ ensures that the function exists and is unique (it also ensures that Eq. (4.3.2) has a solution). Note also that the function $r(z)$ can be chosen arbitrarily so that on the Riemann surface F, the Riemann problem

$$\psi^+(z)=e^{\theta(\pi(z))-\theta(z)}r(z)^{-1}r(\pi(z))\psi^-(z),\qquad z\in\mathcal{L}_0,\ \psi(z_0)=0$$

in the function $\psi(z)$ holomorphic outside the loop $\mathcal{L}_0$ had a unique solution. Solutions to the KP equation obtained in the described fashion naturally coincide with those found by means of the Baker–Akhiezer functions.

APPENDIX

TABLE I

1	$\partial \in \mathrm{Der}(\alpha), \quad \tilde{\partial} = \Gamma_l^{-1}\partial\Gamma_l$	$\tilde{\partial} = \gamma_l\alpha + \partial$
2	$\partial \in \mathrm{Der}(I)$	$\tilde{\partial}(e) = \gamma, \quad \tilde{\partial}^2(e) - \tilde{\partial}(e)\gamma = \partial\gamma,$ $\tilde{\partial}^3(e) - \tilde{\partial}^2(e)\gamma = 2\gamma\partial\gamma + \partial^2\gamma,$ $\tilde{\partial}^4(e) - \tilde{\partial}^3(e)\gamma = 3(\gamma^2\partial\gamma$ $+\gamma\partial^2\gamma + \partial\gamma\partial\gamma) + \partial^3\gamma$
3	$D(\Gamma) = A\Gamma,$ $\partial \in \mathrm{Der}(\alpha), \quad \partial A = 0$	$\tilde{\partial}\widetilde{D}(e) - \widetilde{D}(e)\gamma = 0$
4	$\partial \in \mathrm{Der}(\alpha)$	$\partial(\Gamma^{-1}) = -\gamma\alpha(\Gamma^{-1})$
5	$\partial \in \mathrm{Der}(I)$	$\partial(\Gamma A\Gamma^{-1}) = \Gamma(\partial A + [\gamma, A])\Gamma^{-}1$
6	$\partial_1 \in \mathrm{Der}(\alpha_1),$ $\partial_2 \in \mathrm{Der}(\alpha_2),$ $\partial_1\partial_2 = \partial_2\partial_1$	$\gamma_1\alpha_1(\gamma_2) + \partial_1(\gamma_2)$ $= \gamma_2\alpha_2(\gamma_1) + \partial_2(\gamma_1)$
7	$\partial \in \mathrm{Der}(\alpha), \quad \partial_\beta = \beta - I,$ $\beta \in \mathrm{Aut}(K), \quad \partial\partial_\beta = \partial_\beta\partial,$ $D(\Gamma) = A(\partial_\beta + I)\Gamma, \quad \partial A = 0$	$\tilde{\partial}\widetilde{D}(e) - \widetilde{D}(e)\gamma$ $-\widetilde{D}(e)\partial_\beta(\gamma) = 0$
8	$B \in K, \quad B^2 = e$	$\{B, [B, A]\} \equiv 0, \quad B[B, A] \equiv -[B, A]B,$ $[B, [B, A]] = 2B[B, A]$

TABLE II

$$\partial\Gamma(e-P) = \Gamma N(e-P); \qquad P\gamma P$$

1	$(\partial_0+\partial^3)\Gamma = C\Gamma, \quad (\partial_1+\partial^2)\Gamma = A\Gamma,$ $\partial_0, \partial, \partial_1 \in \mathrm{Der}(I)$	$\partial(4\partial_0\gamma+\partial^3\gamma+6\partial\gamma\partial\gamma)$ $+3\partial_1^2\gamma+6[\partial_1\gamma,\partial\gamma]=0$
1′	$\partial_1 = 0$	$u=\partial\gamma,$ $4\partial_0 u+\partial^3 u+6(u\partial u+\partial u u)=0$
1″	$\partial_0 = \lambda\partial, \quad \lambda\in Z(K)$	$4\lambda\partial^2\gamma+\partial^4\gamma$ $+6(\partial^2\gamma\partial\gamma+\partial\gamma\partial^2\gamma)$ $+3\partial_1^2\gamma+6[\partial_1\ \gamma,\partial\gamma]=0$
2	$(\partial_0+\partial^3)\Gamma = C\Gamma, \quad \partial\Gamma B = A\Gamma,$ $\partial_0, \partial \in \mathrm{Der}(I)$	$u=[\gamma,B],$ $4\partial_0 u+\partial^3 u$ $+3(u^2\partial u+\partial u u^2)=0$
3	$\partial_0\Gamma+\partial^2\Gamma B = C\Gamma,$ $\partial_1\Gamma+\partial\Gamma B = A\Gamma,$ $\partial_0, \partial, \partial_1 \in \mathrm{Der}(I)$	$u=[\gamma,B], \quad v=\{\gamma,B\},$ $2\partial_0 uB+(\partial_1^2+\partial^2)u+2u^3$ $-2\{u,\partial v\}=0,$ $\partial_1 v+\partial v B=u^2$
3′	$\partial_1 = 0$	$u=[\gamma,B],$ $2\partial_0 uB+\partial^2 u+2u^3=0$
4	$\partial\Gamma = (\partial_\alpha+I)^2\Gamma,$ $\partial\Gamma+\Gamma = A(\partial_\alpha+I)\Gamma,$ $\partial_\alpha = \alpha - I, \quad \alpha\in\mathrm{Aut}(K),$ $\partial\in\mathrm{Der}(I)$	$u=e+\partial_\alpha(\gamma),$ $\partial u = u\alpha(u)-\alpha^{-1}(u)u$

$$\partial\Gamma(e-P) = \Gamma N(e-P), \qquad N \in K^{-1}; \qquad PN^{-1}\gamma P$$

5	$\partial_0\Gamma + \partial^2\Gamma B = C\Gamma, \partial\Gamma B = A\Gamma,$ $A \in K^{-1}, \partial_0, \partial \in \mathrm{Der}(I)$	$S = \gamma B\gamma^{-1},$ $4\partial_0 S = [S, \partial^2 S]$
6	$\partial\Gamma = (\partial_\alpha + I)\Gamma,$ $\partial^2\Gamma + \Gamma = A(\partial_\alpha + I)\Gamma,$ $\partial_\alpha = \alpha - I, \alpha \in \mathrm{Aut}(K), \partial \in \mathrm{Der}(I)$	$\partial(\gamma^{-1}\partial\gamma) = \gamma^{-1}\alpha(\gamma) - \alpha^{-1}(\gamma^{-1})\gamma$
7	$\partial_1\partial\Gamma = \Gamma, \quad \partial\Gamma = A(\partial_\alpha + I)\Gamma,$ $A \in K^{-1}, \partial, \partial_1 \in \mathrm{Der}(I),$ $\partial_\alpha = \alpha - I, \alpha \in \mathrm{Aut}(K)$	$\partial_1(\gamma^{-1}\partial\gamma) = \gamma^{-1}\alpha(\gamma) - \alpha^{-1}(\gamma^{-}1)\gamma$
7′	$\alpha^2 = I$	$\partial_1(\gamma^{-1}\partial\gamma) = \gamma^{-1}\alpha(\gamma) - \alpha(\gamma^{-1})\gamma$
8	$\partial\Gamma = (\partial_\alpha + I)^2\Gamma, \partial\Gamma + \Gamma = A(\partial_\alpha + I)\Gamma,$ $\partial_\alpha = \alpha - I, \alpha \in \mathrm{Aut}(K), \partial \in \mathrm{Der}(I)$	$u = \alpha^{-1}(\gamma^{-1})\gamma,$ $\partial u = u\alpha(u) - \alpha^{-1}(u)u$

$$\partial_i\Gamma(e-P) = \Gamma M_i(e-P), \qquad [M_i, \mathcal{D}_i] = 0;$$
$$P\gamma_i P \quad (i = 1, 2)$$

9	$\partial_i\Gamma = (\partial\Gamma + A\Gamma)D_i, D_i \in K^{-1}$ $\partial_1, \partial_2, \partial \in \mathrm{Der}(I), [D_1, D_2] = 0,$ $\partial_1 D_2 = \partial_2 D_1 = \partial D_1 = \partial D_2 = 0$	$\gamma_i = \Gamma^{-1}\partial_i\Gamma, v_i = [\gamma_i, D_1]D_i^{-1},$ $\partial_2 v_1 - \partial_1 v_2 - \partial(v_1 D_2 - v_2 D_1) = [v_1 v_2],$ $[v_1, D_2] = [v_2, D_1]$
9′	$\partial = 0$	$\gamma_i = \Gamma^{-1}\partial_i\Gamma, v_i = [\gamma_i, D_i]D_i^{-1},$ $\partial_2 v_1 - \partial_1 v_2 - [v_1, v_2] = 0,$ $[v_1, D_2] = [v_2, D_1]$

$$\partial_i\Gamma N_i(e-P) = \Gamma(e-P), \qquad [N_i, D_i] = [P, D_i) = 0, \qquad N_i \in K^{-1};$$
$$P\gamma_i N_i P \qquad (i = 1, 2)$$

10	$\partial_i\Gamma = A_i\Gamma D_i, A_i, D_i \in K^{-1},$ $[D_1, D_2] = \partial_1 D_2 = \partial_2 D_1 = 0,$ $A_1 A_2 = \lambda_1 A_1 + \lambda_2 A_2$ $\partial_1, \partial_2 \in \mathrm{Der}(I), \lambda_1, \lambda_2 \in Z(K)$	$\gamma_i = \Gamma^{-1}\partial_i\Gamma, u_i = \gamma_i D_i\gamma_i^{-1},$ $\partial_2 u_1 = -\lambda_1[u_1, u_2],$ $\partial_1 u_2 = -\lambda_2[u_2, u_1]$

REFERENCES

1. Akhiezer, N. I. "Continual analog of orthogonal polynomials on a system of intervals" (in Russian). *Doklady AN SSSR* **141** (1961), 263–266.
2. Dubrovin, B. A., and Novikov, S. P. "Periodic problem for Korteweg–de Vries and Sturm–Liouville equations: their relation with algebraic geometry" (in Russian). *Doklady AN SSSR* **219** (1974), 531–534.
3. Dyn'kin, E. M., and Osilenker, B. P. "Weight estimates of singular integrals and their applications" (in Russian). In *Matematicheskiĭ analiz*, vol. 21 (Itogi nauki i tekhniki VINITI AN SSSR), Moscow, 1983, 42–129.
4. Zakharov, V. E., Manakov, S. V., Novikov, S. P., and Pitayevsky, L. P. *Soliton Theory and Inverse Problem Method* (in Russian). Moscow: Nauka (1980).
5. Zakharov, V. E., and Mikhailov, A. V. "Relativistic-invariant two-dimensional models of field theory integrable by the inverse problem method" (in Russian). *ZhETF* **74** (1978), 1953–1973.
6. Zakharov, V. E., and Faddeev, L. D. "The Korteweg–de Vries equation is a completely integrable Hamiltonian system" (in Russian). *Funktsional'nyĭ analiz* **5** (1971), 18–27.
7. Zakharov, V. E., and Shabat, A. B. "An exact theory of two-dimensional self-focussing in one-dimensional wave automodulation in nonlinear media" (in Russian). *ZhETF* **61** (1971), 118–134.
8. Zakharov, V. E., and Shabat, A. B. "Scheme of integrating nonlinear equations of mathematical physics by the inverse scattering problem method" (in Russian). *Funktsional'nyĭanaliz* **8** (1974), 43–53.
9. Zverovich, E. I. "Boundary problems of analytic function theory in Hölder classes on Riemann surfaces" (in Russian). *Uspekhi Mat. Nauk* **26** (1971), 113–179.
10. Its, A. R. "Canonical systems with finite-zone spectrum and periodic solutions to the nonlinear Schrödinger equation" (in Russian), *Vestnik LGU*, ser. matematika, mekhanika, astronomiya, no. 7 (1976) 39–46.
11. Its, A. R., Kotliarov, V. P. "Explicit formulas for solutions to the nonlinear Schrödinger equation" (in Russian). *Doklady AN USSR*, ser. A, no. 11 (1976), 965–968.
12. Its, A. R., and Matveew, V. B. "On a class of solutions to the Korteweg–de Vries equation" (in Russian). *Problemy matematicheskoi fiziki*, no. 8 (1976), 70–92.
13. Kozel, V. A., and Kotliarov, V. P. "Almost-periodic solutions of equation $u_{tt} - u_{xx} + \sin u = 0$" (in Russian). *Doklady AN USSR*, ser. A, no. 10, (1976), 878–880.
14. Krichever, I. M. "Algebraic geometry methods in nonlinear equation theory" (in Russian). Uspekhi Mat. Nauk **32** (1977), 183–208.
15. Lax, P. D. "Integrals of nonlinear equations of evolution and solitary waves." *Comm. Pure Appl. Math.* **21** (1968), 467–490.

16. Marchenko, V. A. "The periodic Korteweg–de Vries problem" (in Russian). *Doklady AN SSSR* **217** (1974), 276–279.
17. Marchenko, V. A., and Tarapova, E. I. "A new approach to the problem of integrating certain nonlinear equations" (in Russian). *Uspekhi Mat. Nauk* **36** (1981), 227–228.
18. Mikhlin, S. G. *Multidimensional Singular Integrals and Integral Equations.* Moscow: Fizmatgiz (1962).
19. Naymark, M. A. *Normed Rings* (in Russian). Moscow: Nauka (1968).
20. Nikolskiĭ, N. K. *Lectures on Shift Operators* (in Russian). Moscow: Nauka (1980).
21. Novikov, S. P. "The periodic Korteweg–de Vries problem, I" (in Russian). *Funktsional'nyĭ analiz* **8** (1974), 54–66.
22. Tarapova, E. I. "N-soliton solutions to a nonlinear system of equations" (in Russian). *Teoriya funktsyi, funktsyonal'nyĭ analiz i ikh prilozheniya*, no. 36 (1982), 103–111.
23. Tarapova, E. I. "On integrating an n-wave problem" (in Russian). *Ibid.* 38 (1982), 101–103.
24. Tarapova, E. I. "Integration of generalized nonlinear Schrödinger equations" (in Russian). *Ibid.* 42 (1984), 122–131.
25. Tarapova, E. I. "On the integration of certain nonlinear equations in partial derivatives" (in Russian). Thesis, Kharkov State University, Kharkov (1982).
26. Hirota, R. "Direct methods in soliton theory." In *Solitons*, ed. R. K. Bullough, and P. J. Caudrey. Berlin, Heidelberg, New York: Springer-Verlag (1980).
27. Ablowitz, M., and Fokas, A. "Linearization of the Korteweg–de Vries and Painlevé II equations." *Phys. Rev. Lett.* **47** (1981), 1096–1100.
28. Adler, M. "On a trace functional for formal pseudodifferential operators and the symplecture of the Korteweg–de Vries type equations." *Invent. Math.* **50** (1979), 219–248.
29. Gardner, C. "Korteweg–de Vries equation and generalizations, IV: The Korteweg–de Vries equation as a Hamiltonian system." *J. Math. Phys.* **12** (1971), 1548–1551.
30. Gardner, C., Green, G., Kruskal, M., and Miura, R. "A method for solving the Korteweg–de Vries equation." *Phys. Lett.* **19** (1967), 1095–1098.
31. Lax, P. "Periodic solutions of the KdV equation." *Lect. Appl. Math.* **15** (1974), 85–96.
32. Matveew, V. B. "Darbou-invariance and the solutions of Zakharov–Shabat equations." Preprint no. 79/7, Université de Paris-Sud (1979).
33. McKean, H., and Van Moerbeke, P. "The spectrum of Hill's equation." *Invent. Math.* **30** (1975), 217–274.